SIXTY HARVESTS LEFT

HOW TO REACH A NATURE-FRIENDLY FUTURE

PHILIP LYMBERY

BLOOMSBURY PUBLISHING

LONDON · OXFORD · NEW YORK · NEW DELHI · SYDNEY

BLOOMSBURY PUBLISHING
Bloomsbury Publishing Plc
50 Bedford Square, London, WC1B 3DP, UK
29 Earlsfort Terrace, Dublin 2, Ireland

BLOOMSBURY, BLOOMSBURY PUBLISHING and the Diana logo are trademarks of
Bloomsbury Publishing Plc

First published in Great Britain 2022
This edition published 2023

A catalogue record for this book is available from the British Library

ISBN: HB: 978-1-5266-1932-7; TPB: 978-1-5266-1933-4; PB: 978-1-5266-1934-1;
EBOOK: 978-1-5266-1930-3; EPDF: 978-1-5266-5469-4

2 4 6 8 10 9 7 5 3 1

Typeset by Newgen KnowledgeWorks Pvt. Ltd., Chennai, India
Printed and bound in Great Britain by CPI Group (UK) Ltd, Croydon CR0 4YY

To find out more about our authors and books visit www.bloomsbury.com
and sign up for our newsletters

'Intelligent and very well researched' Jamie Blackett, *Daily Telegraph*

'Meticulously researched and engagingly written' Rosemary Miller, *Cape Argus*, South Africa

'The warnings are coming thick and fast, and Lymbery's are clear, concise and truly frightening – we are burning and poisoning the global larder. But we have solutions that we must implement now' Chris Packham

'Philip Lymbery pulls no punches in cataloguing the calamitous mistakes we've made in our food system, but he has bold and inspiring solutions to offer, too. It's time for Big Food, and governments everywhere, to act on them' Hugh Fearnley-Whittingstall

'The chilling title is the red flag; the contents, however, lay out all the remedies to save the planet and its species, including ours, and make for absorbing, sometimes terrifying, reading. Minutely researched, and written for laymen as well as experts, *Sixty Harvests Left* deserves to be read worldwide and acted upon immediately. I cannot recommend it highly enough' Joanna Lumley

'Beautifully crafted. A compelling, excoriating account of industrial farming – how it is driving the climate and biodiversity emergencies, while also undermining our health. Full of insights and encounters with pioneers of new ways of farming, *Sixty Harvests Left* is a call to action – to change our world from the ground up. A vitally necessary book' Isabella Tree

'Thought-provoking … *Sixty Harvests Left* gives us reason to look forward to a brighter farming future and the possibilities that can be achieved through care of our greatest natural asset, soil … Lymbery speaks to practitioners with their feet firmly on the ground and gives hope that new ways in farming will provide for a better future. A fascinating and positive read!' Jake Fiennes

'Powerful, purposeful and persuasive, read Philip Lymbery's book and we know what has to be done. It's simple really, look after the land, farm it sensitively, tread softly on this earth and all can still be well. We need to transform ourselves rapidly. This book is transformative. We must read, mark and learn, fast' Michael Morpurgo

'The true horror story of our current dependence on factory farming and intensive agriculture gets clearer by the day. Philip Lymbery pulls no punches in painting that grim picture… [But] you need to read *Sixty Harvests Left* more for its utterly convincing alternative vision of farming and food production available to us in the near future … I'd be very surprised if you don't end up appreciating this book as much as I did' Jonathon Porritt

'In this beautifully written book, Philip Lymbery describes how intensive agriculture harms the environment and inflicts suffering on sentient animals. But after visiting and talking to those on the front line – scientists, farmers and food providers – he is able to show that there are sustainable alternatives. And that they are working. There is indeed hope for the future of our planet, and each one of us can play a part. I urge you to read *Sixty Harvests Left*' Dr Jane Goodall, DBE, Founder of the Jane Goodall Institute & UN Messenger of Peace

'Not only beautifully written, this is jam-packed with the evidence we need to change our lives in order to save our planet. Philip Lymbery draws us in, in a lyrical and seductive manner, whilst imparting vital, life-changing information. Only we can save our planet and *Sixty Harvests Left* shows us how' Peter Egan

'Philip Lymbery is one of the few who really understand the connections between farming and nature … He is the most important thinker writing about these crucial issues – and the way forward' Carl Safina, author of *Beyond Words*

'An urgent, evidence-based, visionary approach to the most challenging decisions facing humanity. This is a brave, fascinating, game-changing book' Sophie Pavelle

'Excellent – personal and engaging. Lymbery's life experiences make it very readable, allowing him to speak with authority and honesty … An important challenge to the vested interests that make our life on earth unsustainable' Rebecca Nesbit, author of *Tickets for the Ark*

'Philip Lymbery's great service, through beautiful prose and deep research, is to amplify the siren call from our planet and the web of life. Through him, change becomes not only necessary and desirable, but irresistible' Raj Patel, author of *Stuffed and Starved: The Hidden Battle for the World Food System*

'A highly engaging, uplifting story of the author's own experiences with farming and nature, interwoven with detailed research and thorough expertise. Crucially, it ends with hope, recounting the solutions being developed by innovators working to regenerate our soils, and to restore the landscapes dependent on them' Andrew Knight, Professor of Animal Welfare and Ethics, & Founding Director, Centre for Animal Welfare, University of Winchester

PHILIP LYMBERY is Chief Executive of the international farm-animal-welfare organisation Compassion in World Farming. He has played a leading role in many major animal welfare reforms, including Europe-wide bans on veal crates for calves and barren battery cages for laying hens. He was appointed an ambassadorial 'Champion' for the UN Food Systems Summit in 2021. He also spearheaded Compassion in World Farming's engagement with more than 1,000 food companies worldwide, leading to genuine improvements in the lives of more than two billion farm animals every year.

A columnist for the *Scotsman*, his first book, *Farmageddon*, was listed as a Book of the Year by *The Times*, while his second book, *Dead Zone*, was selected as a 'Must Read' by the *Daily Mail*. A visiting professor at the University of Winchester, he is also a keen ornithologist.

To the memory of Rosemary Marshall: 'Onwards and upwards'

CONTENTS

PREFACE

Palm Sunday, 14 April 1935, and a rose-pink dawn broke over Boise City in Oklahoma. There was a freshness in the air, a feeling of spring, a rare sense of relief from the winds that blew relentlessly over the dusty plains of the American Midwest. The sun shone over simple timber homesteads. The blades of a windmill rested motionless beside a pitched roof dotted with pigeons. The sweet whistles of meadowlarks cut through crystal air. It was an unusual calm before an extraordinary storm.

To locals used to life on the harsh plains, it felt like a new beginning. They opened windows and hung out laundry, spruced up their homes and walked to church in the sun. These were tough people who stoically weathered whatever was thrown at them. Yet nothing could have prepared them for what was about to happen.

Out of nowhere, a boiling black cloud as big as a mountain range rolled towards Boise City at frightening speed. Winds of up to sixty-five miles per hour gathered a wall of dust two hundred miles wide and thousands of feet high, all of which bore down on this exposed community. When it arrived, it hit like a brick. Temperatures plummeted and day turned to night, with visibility reduced to zero. Those caught out in the open crawled to find shelter. Cars came to a standstill. The dust made residents choke as it piled up against their homes. One woman thought of killing her child rather than leaving the baby at the mercy of Armageddon.[1] Many people thought it was the Apocalypse.[2]

'It was just like a large rolling pin coming down on you, or a steamroller. I was petrified,' recalled Louise Forester Briggs, a young girl when the storm hit.[3] The legendary folk singer-songwriter Woody Guthrie remembered 'it got so dark that you couldn't see your hand before your face, you couldn't see anybody in the room'. Then twenty-two years old and living in Oklahoma, he saw the storm coming and ran for shelter, later speaking of how people thought it was 'the end of the world'.[4]

The 1930s Dust Bowl was rooted in the unintended consequences of decades of decisions by government policymakers, which struck at the very heart of society. American settlers had migrated westward in search of new lives and prosperity, aided by extraordinary government encouragement; prairieland was offered for free to anyone willing to make a go of self-sufficiency on land once dominated by Native Americans and vast herds of wild bison. Thoughts of producing enough for your family soon turned into aspirations of affluence and making a quick buck. Small-scale farming turned into an industry, with new farming techniques that broke up fragile soils and left them exposed to the elements. Grassland was ploughed for crops that were initially so productive that harvests broke all previous records. Farming became geared towards short-term profit. But when markets were flooded and prices slumped, farming methods intensified; after all, if you could no longer make the same money per acre, then surely the answer lay in farming more acres. However, nature bit back and dust storms blew up, carrying away topsoil, burying crops, flattening livelihoods and shortening lives as people got sick from the dust.

'Black Sunday', as it became known, saw the worst of the dramatic 'dusters', the black clouds carrying millions of tons of topsoil that pounded the Great Plains during the 1930s. The states of Colorado, Kansas, New Mexico, Oklahoma and Texas were particularly hard hit. When the dust reached cities like New York and Washington DC, the enormity of the situation struck home: this was no longer a localised problem.

The Dust Bowl era tested settlers to breaking point. Families struggled to feed their children. While many stuck it out, others fled in search of more favourable conditions, sparking the biggest exodus in American history. In echoes of today's global warming debate, there was resistance to change. Efforts to prevent the plains from being blown away were hampered by arguments that the Dust Bowl was a natural event, that misplaced farming techniques weren't to blame and that things would naturally come right again.

Solving what has since come to be seen as the worst man-made environmental disaster in American history took great resolve and far-sighted leadership. Everyone involved played their part in turning things around, a remarkable example of humans overcoming an environmental monster unleashed by human hand, against the most impossible odds.

History can forewarn us of future dangers. It can also help us gather the people, the ideas and the will to persevere and overcome. Today, there is a gathering storm of monumental proportions: a climate, nature and health emergency. And at its heart is industrial agriculture.

Scientists have warned that we have until 2030 not only to identify solutions to climate change but to make big strides in implementing them worldwide, if we are to 'flatten the curve' of greenhouse gas emissions sufficiently to stop runaway global heating. Food is a major contributor to emissions, yet it is often overlooked. The livestock sector alone produces more greenhouse gases than the direct emissions of the world's planes, trains and cars combined.

At the same time, pollinating insects such as bees that play an essential role in food production are in steep decline. Since 1990, around a quarter of bee species have disappeared,[5] and the losses go much further – in the half-century since the widespread adoption of intensive agriculture, the world has lost 68 per cent of all its wildlife.[6] That's more than two-thirds of the world's wild mammals, fish, birds, reptiles and amphibians – all gone.

I've written this book during the Covid-19 crisis, which has had a profound societal impact. Emergency lockdown measures imposed across the world and a global death toll surpassing 2 million in the first year alone have demonstrated the fragility of our society.

Life-altering changes have largely been seen as beneficial, but the coronavirus pandemic has shown that we can't take progress for granted. Covid-19 is widely seen as a virus that jumped the species barrier into humans, with devastating consequences. Three-quarters of all human diseases originate from animals. A decade earlier, swine flu did the same thing: having originated in pig factory farms, it caused around half a million deaths worldwide. But the link between disease and farm animals doesn't end there. Nearly three-quarters of all the world's antibiotics are fed to farmed animals, largely to control disease inherent in the cramped, squalid conditions of animal factories. This is a major reason for the World Health Organization warning that action is needed if we are to avoid a post-antibiotic era, where diseases that are currently treatable could once again kill. Experts have suggested that by the middle of the century, the demise of antibiotics could lead to 10 million deaths a year.[7]

The common thread in all these challenges is industrial agriculture, where animals are reared in confinement, while crops are grown in huge, prairie-like fields of a single variety using chemical pesticides and artificial fertilisers, at great cost to our health, the environment and animal welfare. Every year, 80 billion chickens, pigs, cows and other farmed animals are produced worldwide, with two-thirds reared using intensive or industrial agriculture, also known as 'factory farming'. Caged chickens have so little space that they are unable to flap their wings. Pigs are kept in narrow crates or crowded pens, unable to nuzzle their snouts in the dirt. And cows stand listlessly in crowded feedlots or indoor dairies instead of grazing in green fields. Fish are also industrially reared, with half of the world's supply coming from farms. The irony is that we continue to plunder the oceans for fish to feed to these farmed fish, a huge problem as fears mount that overfishing could cause the collapse of wild fish stocks.[8]

And the end result of factory farming is that meat, and the land and aquatic animals that produce it, have come to be seen as disposable. UN estimates suggest that the amount of meat wasted every year is equivalent to 15 billion animals being reared, slaughtered and binned.[9]

Industrial agriculture has its origins in the Dust Bowl era, when soil was seen as 'inexhaustible'. Crop growing turned into factory farming, with agricultural holdings increasingly seen as food 'factories'. Crops and animal products began to achieve record yields, creating a surplus that exceeded demand, causing prices to slump. The Wall Street Crash in 1929 dragged the US into the Great Depression. Desperate times led to desperate measures. With prices cut to the bone, cash-strapped farmers ploughed more pasture – if they couldn't get as much money for their produce, they'd grow more to make up for it. It was a vicious cycle. In the mid-1930s, the US government stepped in with subsidies aimed at keeping farming afloat.

An unforeseen consequence of the slump in prices was the use of cereals as animal feed, a practice that decades of government subsidies has since perpetuated; factory farming, the feeding of confined animals with grain, started to spread around the world. After the Second World War, munitions factories were recommissioned to produce fertiliser to boost grain production, while chemical weapons were repurposed as chemical pesticides. Post-war aid helped spread the means of intensive farming first into Europe and then beyond, and a whole industry grew up around it – fertiliser and pesticide companies, cage manufacturers, companies selling specialist breeding stock and even drug companies that targeted factory-farmed animals as a market for antibiotics. Production became bigger and more centralised. Factory farming was born.

In little more than half a century, factory farming has become the biggest global cause of animal cruelty and wildlife decline. As well as contributing to biodiversity loss, climate change and squandering antibiotics, industrial animal farming even affects oceans. About a fifth of the world's fish catch is used in animal feed. Pollution from expanding intensive croplands has also caused hundreds of oceanic dead zones, one of the biggest being an area of polluted sea the size of Wales in the Gulf of Mexico.

This intensification of food production may seem like a space-saving idea, with animals caged and confined on industrial farms, but actually isn't – vast acreages of arable land have to be devoted to growing the feed. 40 per cent of the global grain harvest,[10] enough food to sustain 4 billion people[11], is used as animal feed, causing much of its calories and protein to be lost. If all the grain that is used for animal feed were grown in a single field, it would cover the entire land surface of the UK and EU put together.

Far from sparing land for nature, the reality of intensive farming is that nature is wiped away while farmland continues to expand. An increasing global appetite for meat from factory farms means that more forests are cleared for farmland, encroaching on wild lands. Our diet has become imbalanced and over-dependent on meat and dairy, putting ecosystems under greater pressure.

Yet there is another way. Moving from industrial agriculture gives us the opportunity to put decent food on our plates and save the natural world. The key to that better way can be found on the world's pastures, where we graze animals instead of feeding them grain. Compared to pasture-raised beef or free-range chicken, the factory-farmed equivalent has as much as twice the saturated fat and is lower in other nutrients.

Keeping animals on the land regeneratively in mixed rotational systems, with pigs and chickens fed on crop residues and food waste, provides a more efficient way of producing nutritious, nature-friendly food, while also providing scope for the highest standards of animal welfare. As well as restoring farmed animals to the land, it is important that we reduce their numbers too. In order to stabilise the climate and save the natural world, we must reduce global meat and dairy production by at least half in the next thirty years. If we carry on eating meat and dairy as we are, our food alone could trigger catastrophic climate change. But a smaller, more climate-friendly livestock population can be raised on the world's mixed farms and pastures, using ruminants like cattle and sheep as grazing animals and pigs and chickens as recyclers of food waste and farming's leftovers. Rewilding the soil has a big part to play in ensuring future harvests, bringing back the great weight of biodiversity beneath our feet which helps store carbon and water naturally. A handful of healthy soil harbours more living organisms than the number of people on the planet. As we'll discover, restoring this richness not only preserves our ability to grow food in the future but can also lead to more nutritious food for all.

Through my work with the animal welfare charity Compassion in World Farming, I have learned that treating animals with compassion and respect is at the heart of sustainable food. I've never forgotten the lessons of the charity's founder Peter Roberts, a dairy farmer who saw that improving animal welfare is an essential part of protecting the environment.

Peter was ahead of his time in identifying the connections between food, farm animals, wildlife, plants and the soil. I remember a seminal leaflet of his, 'Aims and Ambitions', in which he described why it was important to move away from diets heavy in animal products and particularly those derived from grain-guzzling chickens and pigs, who consume food that might be fed to people directly. 'We must progressively reduce our dependence on animals and gradually relegate them to the less productive land. As we do so we must shift the emphasis away from monogastric animals', he wrote. I always felt Peter should write a book spelling out the connections between farm animals and nature, the soil and us, but he never did. As his protégé, thirty years on, this is that book. Sitting alongside my other books inspired by him, *Farmageddon* and *Dead Zone*, I hope it will offer solutions to some of the world's most pressing problems.

In recent years, I have had the privilege of travelling the world and have seen for myself what the industrialisation of agriculture has done to animal welfare, the environment and our health. I've spent time in the rainforests of South America and Sumatra, been to the breadbaskets of Europe, the UK and the USA, and immersed myself in the agricultural realities of China and Africa. I have poured these experiences into this, the third book of the *Farmageddon* trilogy, much of this one having been written during the Covid-19 lockdown on the farm hamlet where I live.

My sense of the connectedness of food, farm animals and nature has only been heightened by living on a farm. Every day, I see the good and bad in farming. The sight of cows grazing in valley meadows brings me joy. Seeing arable soil running into rivers and onto roads makes my heart sink. At the heart of this book is the secret of how we can reconnect our food to the soil and get farmed animals back to the land where they can enjoy fresh air, sunshine and the earth under their feet. There should be fewer of them, kept better. In short, a decent future for people and animals, both farmed and wild, relies on the regeneration of the soil.

FROM NOMADS TO SETTLERS

Ten thousand years ago, our ancestors changed the course of human history: they became settlers. Until then, *Homo sapiens* were predominantly nomadic hunter-gatherers, powered by the natural bounty around them and their ability to catch or find it, but also limited by it.

As nomadic lifestyles gave way to settlement, the age of agriculture was born. Before then, if the ecosystem we were part of didn't provide, we moved elsewhere. The hand-to-mouth existence of finding sustenance for today and then doing the same thing tomorrow gave way to investing in the future through the land, creating our own luck through planting, grazing and tillage. And that investment allowed the development of kingdoms and nations, taxes and courts, culture, reading and religion.

Becoming settlers saw us create a contract with the soil, which became the bedrock upon which everything is founded. We invested in it, planting the seeds of future harvests and grazing our animals in ways that would nurture our prospects of a decent return. What we now call the 'nitrogen cycle' became the most fundamental example of the 'circular economy'. Sunlight is captured by plants and converted into things that we or our farm animals eat, with the waste returning to the ground in the form of compost and manure and replenishing the soil for future harvests.

For millennia, the soil provided the food that allowed civilisation to flourish. But a single human lifetime ago, we started to break the bond, to tear up that contract with the soil, when we lost sight of food and soil as part of nature. We started to remove farm animals from the land and place them in confinement, regarding animals as machines and the soil as little more than a growing medium. Instead of returning organic matter to the soil, we applied chemical fertilisers, which fed the plant but undermined the soil as a living ecosystem, with far-reaching consequences. This new approach to farming led to a huge leap in food production, but it could well be the undoing of us. As the UN Food and Agriculture Organization puts it, 'this intensive crop production has depleted the soil, jeopardizing our ability to maintain production ... in the future'.[12]

Soils are ebbing away so fast that they could be useless or gone in a lifetime. And then what? No soil, no food, game over. According to the UN, if we carry on as we are, there could be just sixty harvests left in the world's soils. Time is running out.

Whether it be fruit, vegetables, cereals or meat and dairy, about 95 per cent of our food is dependent on the soil.[13] Soil captures rainwater too, and holds it against gravity in a way that makes it accessible to plants. Without soil, much rainwater would simply disappear, making its way back to the sea. Soil is also a massive store of carbon that would otherwise be heating the atmosphere. The first thirty centimetres of soil contains around 680 billion tonnes of carbon – almost double the amount present in our atmosphere.[14]

As much as 37 per cent of greenhouse gas emissions globally are caused by our food and the way we produce it.[15] The majority of this comes from agriculture, as well as deforestation to make way for new farmland. Farming already covers half the habitable land surface of the planet; as it has intensified, problems relating to soil, wildlife and animal welfare have mounted. It also accelerates the loss of carbon into the atmosphere from the soil. The world's breadbaskets are at risk, both from soil decline and from global warming. Experts warn of multiple breadbasket failures leading to global food shortages.[16] In the battle to avoid catastrophic climate change, the vast footprint of food and farming creates both a massive barrier and a huge opportunity. Switching to soil-enhancing regenerative and agroecological farming, using techniques that replenish soil fertility and capture carbon along the way, can play a big part.

For decades, food and farming have existed as if in an endless summer, with finite resources used as if they were inexhaustible.

Each part of this book begins with my own observations on the farm where I live, as the changing seasons reflect the development of modern-day agriculture, and the challenges it poses. Summer, a time of consumption without limits, draws to a close as we recognise the consequences of living beyond our planetary boundaries. The climate crisis and the collapse of nature have a feeling of a looming autumn, while Covid-19 has given us a collective taste of winter. If we are to enjoy a new spring, the way we produce food needs to change.

History tells us that major shifts in food production can happen fast but turning around the impending catastrophe will need to be faster still. Within decades, industrial agriculture became a global practice. Changing the way we produce food has never been more urgent, but the global response to the Covid-19 pandemic has shown us that when there's a big enough reason, big change can be effected.

To move as fast as we need to, we are likely to require a multiplicity of approaches rather than a single solution. We will need to fuse regenerative farming with a reduction in animal-sourced foods in ways that will help rewild the soil for future harvests. In this book, we will meet the pioneers, the rebels and the revolutionaries who are creating the changes needed to take us into this new food era. Farmers who farm regeneratively, building healthy soils and allowing nature to thrive, ensuring there will be many more harvests. Food innovators: the scientists who create meat without animals by using cell cultures, and the urban vertical farmers who grow food on a fraction of the land used traditionally. The fermenters who build on long-established methods of producing bread and wine but with a modern twist, creating precise proteins that hold previously unimaginable possibilities. Much of this blends traditional wisdom with new perspectives, while the rest uses technology that is mind-blowingly futuristic.

This is a book is about the opportunities that exist for a rebirth in the countryside. Our future depends on a thriving ecosystem, and that starts with soil.

PART ONE

SUMMER

Early morning in midsummer is a magical time of year in the English countryside. In the warm calm before the start of the day, tall grasses stand statuesque. Without the faintest quiver of breeze, sounds and the senses are heightened. A bee buzzes and a robin sings his silvery trill. The murmur of the river is amplified by the stillness, its flow drifting through the stone arches of a sixteenth-century bridge. A pair of kestrels, slender with long tails and pointed wings, carve silhouettes across the reddening sky. The early glow of sun flashes against three mute swans in flight, their whistling wingbeats resonating in the air. Streaks of cloud turn ochre and then white. In a distant corner of the valley, I can hear our neighbour's cows.

The moment before dawn breaks is the perfect time to gather thoughts. It is also the time when I feel most alive, attuned to my surroundings, connected to them by a sense of anticipation.

I live on a small farm hamlet overlooking the River Rother in a gently undulating part of West Sussex. Ours is the old coaching house, a building that has stood for four centuries. It once served as a dairy. Cows were milked and calves were bedded, while pigs rooted around in what is now our garden.

Here in the South Downs National Park, distant hills fringe the skyline. The nearby river rises out of chalk and clay before weaving through our valley along tree-lined banks of oak, hazel and willow. If I wait patiently on the bridge over the slender waterway, I might catch a glimpse of a dazzling blue kingfisher. Nearby, badgers make a sett in sandy soil, while roe deer hide in thickets and copses. They sometimes stand for a moment with staring eyes, stunted horns and flicking ears, before disappearing into cover. Fields of maize, pumpkins and wheat are lined with hawthorn, holly and blackberry.

From April to November, forty cattle roam the pastures. This is rough grazing, the rich grassland infused with spiny tussocks of thigh-high soft rush. The cows cross the river at will, gathering at favoured shallows before wading together in a line.

The plants and wildlife here have long attracted attention: former local residents include the visionary food writer Patience Gray, who foraged for wild plants and fungi to supplement her cooking during the Second World War. More recently, there was a field centre linked to King's College London, where many budding naturalists, including the TV presenter Chris Packham, learned their subject. Local mice and voles apparently became so accustomed to being live-trapped and studied that they learned to put up with temporary incarceration in return for a free meal. The field centre has since been converted into desirable housing, yet the wildlife still makes a living around here. At least, for now.

I love living on a farm. I've always wanted to be immersed in the folds of a rural setting, a place where things look different each and every day. There's always something new to catch the eye. This is the lens through which I view the countryside. Daily, I walk the fields and woods with our rescue dog, Duke, a large, black crossbreed reminiscent of a baby bear. He came to us as a ten-week-old puppy, having been abandoned by his owner. I love his company and he loves our walks in the countryside that serves as his playground and my anchor.

Living here, I see the good and the bad of farming first-hand. I see the changes in the wildlife: how skylarks have all but disappeared from farmland and how native grey partridges are rare among the many red-legged partridges introduced for the local shoot. I also see soils on the move – those in the river's catchment area are some of the most erodible in the country, yet I still see maize, a common animal-feed crop, planted in tramlines down local hillsides, encouraging soil to run towards the river. And when soil goes into the river, so do fertilisers and pesticides. *After heavy rain, soil can be seen on the roads, which can resemble mudbanks, and in the fields, where deep gulleys open up. The occupants of one hamlet not so far from here woke up one morning to find that a field had slid into their garden, potatoes and all.*

That said, there is still much to celebrate in the countryside, including the return of the red kite and the buzzard, two magnificent species that were once all but eradicated. I can still find orchids in the pasture and at the edges of woodlands. Most years, bramblings and woodlarks visit. Tiny harvest mice live among the lupins near a picturesque old church. Yet there is no escaping the sense that our society seems to live as if in an endless summer, with lifestyles based on short-term thinking and a belief that things will last forever.

As Duke and I shelter from a late-summer downpour in the cow barn next to our house, the sight of soil running towards the river gives me pause for thought. For how much longer can we go on like this?

1

BLACK GOLD

BREADBASKET PIONEERS

Deep in the Fens, Britain's eastern breadbasket, in February 2012, and a lone ploughman was about to make an astonishing discovery. His tractor ploughed perfectly straight furrows across a spirit-level flat field. It tilled back and forth between regimented poplars, planted to prevent soil being carried away by the wind. The steel blades of the plough sliced deep and laid jet-black peat over the surface. Soon potatoes would grow here, shining ivory white in black soil. To the ploughman, his work was unremarkable. He carried out his seasonal task as seamlessly as the warmth of the sun moved through big Fenland skies. But then his plough snagged suddenly, bringing the tractor to a juddering halt. Below the surface, blades had struck something big and immovable. Buried in a peat grave were the perfectly preserved remains of a tree, the likes of which had not been seen for millennia.

An ancient trunk of black oak, forty feet long and more than 5,000 years old, lay within the soil of that Fenland field, a mighty remnant of the dense forest that once covered this flat, now largely treeless region. Described as the biggest piece of 'bog oak' ever found, it was chopped up by a sawmill flown over from Canada to produce what were said to be some of the most valuable planks of wood in the world. They would be crafted into a magnificent table that would grace nearby Ely Cathedral.

Fenland farmers were used to finding bits of bog oak in their fields, but the piece at Wissington Farm was remarkable for its size and preservation. It was thought to have fallen during the Stone Age, at the very birth of farming in Britain. For five millennia, it had lain in an airless grave of peat, preserved by the lack of oxygen.[1] Rising sea levels flooded its roots and those of its neighbours, bringing down what was once a thriving ancient forest.

The farmers behind the discovery drafted in the furniture maker Hamish Low to turn this extraordinary piece of wood into a masterpiece. Low said of the project, 'This one is so special … Along with the fact that it is impossible to know how long Fenland black oaks will continue to rise out of the soil, and their inherent fragility, this one is worthy of preserving for the interest of the nation.'[2]

'Inherent fragility' might also be a description of the Fens themselves and the extraordinary farming that goes on there. For much of the 5,000 years since that bog oak fell, the low-lying Fens were largely wetland. Growing crops was restricted to the few hills known as 'fen-islands', while grazing animals would have taken advantage of seasonal grasslands. Although farming in the Fens dates back to Roman times, the land remained largely underwater until the seventeenth century, when the Dutch engineer Cornelius Vermuyden led the work of pumping water away from the wetland landscape. This engineering feat intensified in the late eighteenth and early nineteenth centuries, when drainage work produced the featureless landscape we know today.

The Fens region is one of the most productive farmlands on the planet, and also one of the most threatened.

Draining the Fens revealed peat soils so rich in nutrients that they became known as 'black gold'. Fertile Fenland soils account for about half of the most productive 'grade one' farmland in England. Covering a million acres, this beating heart of British farming produces a third of English vegetables, a fifth of its potatoes and a big proportion of its cereals.[3] However, rising sea levels present a serious risk to this low-lying coastal plain, which stretches from Cambridge to Lincoln. Set against a steady decline in British food self-sufficiency, the Fens are a precious resource that is pivotal to the nation's future food security.[4] As well as being under growing pressure from sea-level rise, the region is seeing its soils disappearing at an alarming rate.

'We're losing up to half an inch of peat soil a year', said Charles Shropshire, the Fenland farmer whose family's land hosted the bog oak discovery. 'How are we going to do this [farming] when the peat soil is gone?'

In the same year as the Wissington bog oak find, the Shropshire family embarked on a regenerative farming project, a huge undertaking across 13,000 hectares of land, on which they grow three-quarters of the celery and radishes sold in British supermarkets, as well as two-thirds of the beetroot and nearly half the lettuce. Established by Shropshire's grandfather Guy in 1952, G's Fresh has grown into one of the UK's biggest produce companies, supplying all of Britain's major retailers. The numbers are mindboggling: the company turns out a billion packs of produce every year, creating a revenue of £500 million. G's Fresh is a major employer, with 8,000 employees in Britain, the Czech Republic, Poland, Spain and Senegal. Their international operations ensure that supermarkets can stock their fresh salad all year round.

With such a major company at stake, the Shropshire family's foray into regenerative agriculture is led not by ideology but by necessity. As Shropshire told me, 'The next generation will have the full effect of lost peat … we are changing the way we farm to slow this down.' The finite nature of the resource is now widely recognised. 'Fen blows' are common in spring and autumn, when the soil is bare but recently cultivated. The wind gets up across the flat, open expanses and collects a dark cloud of dry topsoil. In scenes reminiscent of the American Dust Bowl, dust blocks out the sun and cars turn on their headlights, as peat dust, grit, grain and fertiliser pellets fall like a hailstorm.[5] One Fenland worker described it to me as being 'like a scene from Armageddon'.

Exposed soils also wash into waterways, get flushed out to sea, or oxidise and disappear into the air. Peat that remained perfectly preserved when submerged without oxygen disappears when drained, releasing carbon dioxide into the atmosphere and contributing to climate change. Holme Fen is home to one of the most potent symbols of soil disappearance. In 1851, when the surrounding wetland was drained, a cast-iron column was driven into the peat with only its top poking out to measure what was happening to the soil. The landowner realised that during the drainage process the peat would shrink, leaving the post to demonstrate how far the land had receded. Today, the top of the pillar sits four metres above ground.[6]

As UK Secretary of State for Environment, Food and Rural Affairs in 2017, Michael Gove cited industrial farming methods, with their 'determination to drive up yields' as causing soils to degrade and become less productive. In a speech at World Wide Fund for Nature's (WWF's) UK headquarters, he said, 'It is not only less effective at sequestrating carbon, it is progressively less fertile. The effect is most noticeable in what has been some of our most fertile growing soil, in the Fens.'[7]

Much of the remaining peat in the Fens is less than a metre deep, and the fertile covering that buried the Wissington bog oak is being lost at the rate of nearly an inch a year.[8] At that rate, it could be gone in less than fifty years. The challenge is that soil is commonly seen by intensive farmers as a growing medium rather than a living organism.

'Soil is at the heart of every decision we make on the farm', Shropshire told me as he beckoned me into a claret Land Rover Discovery. I was visiting in the middle of the Covid-19 crisis, and due to lockdown restrictions, this was my first time away from home in thirteen weeks.

Like many farmers, the Shropshires are finding that degrading soils isn't their only problem: their yields are plateauing, production costs are rising and financial rewards have dropped. 'Instead of sitting around moaning, we're doing something about it', Shropshire said. 'What are we doing? More for less; more from the soils by putting less fertilisers on.'

Among the leading regenerative pioneers in Britain, their aim is to use artificial fertilisers only as a last resort and to restore farm animals to the land in ways that enhance animal welfare and soil fertility. They also plan to stop ploughing altogether, as it disturbs the soil ecosystem and releases carbon into the atmosphere, neither of which is good for sustainability. They have already reduced their use of artificial pesticides, cut ploughing by a third and now protect the soil by keeping it covered by planting cover crops.

In the Fens, like many parts of the countryside, industrialisation has separated farmed animals from the otherwise nature-friendly mixed farm. Much animal farming today is carried out on factory-farms, where chickens, pigs and cows are kept confined in cages, barns and feedlots where they can no longer find their own food. Huge resources are needed to grow their feed, intensifying the strain on the planet. This sort of farming also prevents animals playing their part in the natural agricultural ecosystem, fertilising the ground with their manure and contributing to soil-preserving changes in land use brought on by crop and animal rotation.

Regenerative farming on the other hand aims to farm more in harmony with nature, taking care of precious resources and integrating plants and animals with the land in ways that help to restore nature's resources. By rebuilding nature's soil, carbon, water and wildlife resources, the aim is to go beyond sustainability – being able to do tomorrow what we do today – and to increase the ability to do *more* tomorrow. They plan to do this by avoiding the degradation of land, by having a diversity of plants and animals and rotating them, so that each field regularly has a different crop. And so that soil-fertility-building grasses are grazed or foraged by animals that can enjoy fresh air and freedom.

Re-introducing free-roaming animals is a big part of Shropshire's plan to preserve soils for future generations. 'We're really keen to get diversity back on the land', he explained. Three and a half thousand sheep now graze the cover crops on the family's farm, protecting the soil in winter but organic beef cattle are next on the wish list. Shropshire sees value in them grazing the soil-building pastures that move around the farm, alternating with salads and vegetables in rotation. Grazing maximises the use of the land and their muck naturally returns nutrients to the soil, which is important for the land's future productivity. Chickens are also in the family's sights; Shropshire mentioned the possibility of hens in 'motorhomes' being towed around the fields behind the cattle and sheep. Bringing farm animals back to the land is a way of adding diversity to solve the soil crisis. It does right by future harvests, and also by animals.

For this farming family, bringing back diversity also means restoring wildlife. 'We're aiming for first-class conservation', Shropshire explained. He wants to return farmland bird numbers to pre-1970s levels, the era that most commentators identify as the beginning of a steep decline due to agricultural intensification.

The Shropshires' ethos involves farming the best land and devoting less productive areas to 'rewilding'. They have already been nationally recognised for their efforts to bring back turtle doves, a migratory species that has declined by 98 per cent since 1970 and is one of Britain's most endangered birds.[9] That decline has largely been due to the loss of mixed farming and the increased use of herbicides that remove the weeds and thereby the seeds that doves feed on.

Supported by the Royal Society for the Protection of Birds (RSPB), the Shropshires have set up a reserve of more than ten hectares – roughly ten football pitches – which is thought to be the biggest dedicated farmland space for turtle doves in the country. They aim to satellite-tag turtle doves in the Fens and track them on their 3,000-mile journey to Africa for winter. They hope to lure the doves to the Shropshires' farms in Senegal along the way, making for an international farming and conservation success story.[10]

However, not everything has gone to plan: the doves have so far preferred to nest on a scruffy part of the farm some miles away from the reserve. A few more seasons of rewilding might change that, but in the meantime, endangered farmland birds like corn buntings, skylarks and nightingales are taking advantage of the reserve.

I discussed with Shropshire what it means to be regenerative. We talked about the importance of getting off the chemical treadmill, of rebuilding soil and wildlife, of carbon sequestration and of controlling pests and disease naturally. He cites *Dirt to Soil* by the American pioneer Gabe Brown as a big motivator: 'That book made me believe that it's all possible.'

Shropshire, who is in his early thirties, intends the farm to be fully regenerative within a decade, thanks to what he calls his 'future farming programme'. His father John, chairman of the family business, hosted us for afternoon tea. The company has clearly come a long way in its seventy years. A large painting of the interior of Ely Cathedral dominated the living-room wall. Before it stood a table made from a 5,000-year-old bog oak similar to the one found by the ploughman in Wissington. Sweeping windows looked out over the Fens to where Roman settlers are thought to have grazed their cattle.

Father and son showed a clear understanding that, in order for there to be a decent future, things have to change. This Fenland farming family had an appreciation of their land's history and a keen eye on its future.

RETURN OF THE CRANE

As I shivered in the fading East Anglian light, the wait seemed eternal. Everywhere was cold and quiet, the kind of stillness where the snap of a twig rippled through the brittle air. For ages, nothing happened. Then came an evocative sound that gave me goosebumps: a rattling bugle that recalled the trumpeting of an elephant. Though I was far from Africa, the creatures making this sound were no less exotic. This was what I'd come here for: the sight and sound of a flight of cranes.

Flying fast, with long, deeply fingered wings, protruding necks and trailing legs, they made for an awe-inspiring sight. And when they came close enough, I could make out the black bibs and red helmets of the otherwise silvery adults.

For many years, I made a pilgrimage on New Year's Day to see common cranes returning to roost in the Norfolk Broads. Each time it would be the crowning glory of a day spent marvelling at the rich fauna that survives in the marshes, broads and fens of Britain's easterly coastline.

Centuries ago, these regal birds used to be a common sight in the British countryside. Many place names, such as Cranleigh, Cranfield and Cranmere, reflect the presence of these birds across rivers, farms and lakes, but they were wiped out in Britain during the seventeenth century when their wetland habitats were drained, largely to make way for agriculture.

The Fens were the last stronghold of this charismatic metre-high wetland bird, before it became extinct as a nesting species. I remember being a wildlife-obsessed teenager and hearing older naturalists talking in hushed tones about cranes in Britain making a comeback. I was a regular at a local reservoir where birdwatchers would whisper dramatically, 'They're back!' They were referring to what was, at the time, the worst-kept secret in wildlife: a handful of wild cranes had returned to the Norfolk Broads.[11]

When I actually saw a crane some years later, it was over a hundred miles way away from Norfolk. 'My' bird was quite possibly a lost migrant. I can remember the moment like it was yesterday. I was out for a wildlife foray near London, when a large, majestic bird, all wings, legs and neck, suddenly loomed into view. I could scarcely believe that I was looking at a bona fide British rarity – I had *found* a crane! This was a new experience on two fronts; never having seen a crane before and never having found what *British Birds*, the ornithological magazine that vetted sightings of rare birds, classified as a rarity. For reasons that now escape me, I never sent in my record of that sighting – someone else did, and their name, not mine, stands beside that official record. Why didn't I do it? Maybe I was daunted by the paperwork. Maybe I couldn't be bothered or forgot. Or maybe I was overly optimistic, thinking that discovering rarities was something I was going to do a lot of. Well, thirty-five years later I've managed to find three more, one every ten years. And each time since, yes, I've sent in the paperwork.

Those early experiences instilled in me a real sense of affinity with cranes, those fascinating, almost mythical creatures that represent a thriving countryside. And in those early days, I was aware of how the countryside was declining with the intensification of farming. Not just in and around farmland fields themselves, but in the wider natural environment.

One of the authors on my bookshelves from a young age was Professor Ian Newton, and I was recently struck by the opening words of his latest publication, *Farming and Birds*: 'The effects of modern agricultural practices on the countryside and its wildlife are strikingly evident to every naturalist.' In his book, Newton gives an insightful summary of the 'revolution' in agriculture in the latter half of the twentieth century, and its 'heavy dependence' on artificial fertilisers and pesticides. Farming became more mechanised and large-scale, boosting output beyond what was previously thought possible. However, this narrow focus on production was achieved at 'huge financial and environmental costs, one of which was a massive loss of wildlife, including birds'.[12]

I wanted to find out whether a decline in wildlife is recognised as a bellwether for what is happening out on the farmland that covers much of the country, and indeed the world. For decades, farming has produced plentiful food while destroying wildlife and degrading the land, providing for today at the expense of tomorrow. This tale of demise involves animal and plant producers, both of whom rely on one supremely non-renewable resource: the soil. I was keen to find out from farmers and conservationists how the links between farm animals, wildlife and farming have been playing out in one of Britain's most important food-producing regions.

While our supermarket shelves are well stocked today, this might not always be the case, with farm intensification storing up problems for tomorrow. In fact, empty shelves during the Covid-19 crisis showed just how fragile our way of life really is and that tomorrow's problems are happening now. This simple reality is driving a growing scepticism at the idea that the world can be fed by farming more intensively.

The origins of the problem lie in the mid-twentieth century, when farming was encouraged to scale up and intensify and the long-held cycle of rotating crops with farmed animals was abandoned.

As John Shropshire told me, 'My father's generation changed agriculture. They introduced tractors, pesticides, artificial fertilisers. They introduced new genetics, and actually produced a huge amount more food. My generation only made it more efficient and bigger ... my generation really have taken from the soil,' he said. As for his sons' generation: 'They are now rebuilding the soil.'

There is no doubt that industrial farming has boosted yields beyond all recognition. A farmer in the 1940s would typically get two tonnes of wheat per hectare. The application of chemical pesticides and fertilisers, together with modern crop breeds, saw productivity increase two- or three-fold between 1950 and the 1990s.[13] A farmer in Northumberland recently achieved a record-breaking sixteen tonnes per hectare.[14]

However, the odd remarkable harvest aside, yields in the UK have plateaued since the turn of the century.[15] This is as much due to demise of the natural resource on which farming relies as to any limitation of technology. Chemicals designed to kill undesirable wildlife can also be harmful to earthworms and other essential soil life such as mycorrhizal fungi, the biological network in the soil that helps bring water and nutrients to plant roots. Overuse of these chemicals, along with the mania for stripping out hedgerows that provide food and shelter for many animals, has led to a calamitous loss of wildlife. More than 40 per cent of species in Britain have declined since 1970, while one in every seven of its wildlife species faces extinction.[16] Britain has since become one of the most severely nature-depleted countries on the face of the Earth.

Then there is farming's contribution to climate change. British agriculture is responsible for 10 per cent of the country's greenhouse gas emissions, mainly through methane from cows and sheep, nitrous oxide produced by fertilisers and carbon dioxide produced when carbon-rich organic matter in the soil oxidises during ploughing.[17]

BACK IN THE FENS

The bleak, featureless topography of the Fens is strangely beautiful. All it takes is a sunset and some mist and you're transported back to a medieval landscape. When the late botanist Sir Harry Godwin arrived from Yorkshire and said, 'My God, it's flat here, isn't it?' a local replied, 'Aye, but any fool can appreciate a mountain. It takes of man of discernment to appreciate the Fens.'[18]

Covering the counties of Cambridgeshire, Lincolnshire and Norfolk, the area has a rich ecology, thanks to its nutrient-rich wetlands. An elaborate system of drainage channels and man-made rivers, dykes and drains, driven by automated pumping stations, carries the water uphill and out to sea. Internal drainage boards have been established to maintain 3,800 miles of watercourses and nearly three hundred pumping stations, which have the capacity to pump the equivalent of 16,500 Olympic-sized swimming pools in a day. Holme Fen is renowned for being Britain's lowest point, nine feet below sea level. And the land continues to sink, leaving its future dependent on dykes built to protect it from flooding.[19] And farm animals that once grazed the low-lying grasslands have since been largely replaced by specialist arable cultivation.

It was here that I met Ian Rotherham, professor of environmental geography at Sheffield Hallam University and a prolific writer on the Fens. He grew up passionate about wildlife and, like me, was a keen teenage birdwatcher. He took me to see a classic example of Fenland farming. We drove to Bardney, a village near Lincoln on the east bank of the River Witham, the kind of town where everything stops at 2.30 p.m. With a high street comprising two pubs, a chippy, a butcher's and a place selling guns, it has a long agricultural heritage. That said, things have changed. The skyline as you approach the village is framed by the huge silos of a British Sugar factory, long since closed.

The airfield just outside the village was home to the Royal Air Force during the Second World War, before becoming a base in the Cold War for Thor missiles, the first ballistic missiles to be tipped with nuclear warheads.[20] Now the airfield houses rows of intensive chicken sheds.

Rotherham and I talked beside a road that stood two metres above the surrounding land. There were farm buildings, some of which were derelict. 'Draining an organic peat fen, the soil starts to break down, dries up, so the soil then starts to oxidise, and it basically shrinks before your very eyes. That's why all the land has gone down', he said. 'The soil is the most precious thing that you have: it's your future.'

Rotherham told me that overfarming can cause desert landscapes. 'If you burrow down through the sand, you'll find a civilisation that overreached its carrying capacity. Libya was the breadbasket of the Roman Empire. They were exporting grain across the Mediterranean to the Italians. Look at it now, it's a desert.'

According to Rotherham, Fenland peat soils might have thirty to fifty harvests left before they are gone. Silt fens may have more longevity, but the most extensive peat areas are disappearing at a rate of knots. 'You drive along Cambridgeshire or Lincolnshire in the south and you are up on the road and the land is down there. That's all peat that's shrunk, washed away or blown away,' he said.

Where did it all go wrong? Rotherham identifies the move away from mixed farming as one of the biggest mistakes in recent times. 'It's common sense, but we've kind of broken away from that because of the economics of short-term intensification.' And if things carry on as they are, what will this productive Fenland landscape be like in fifty years' time? 'I think it will be in a very sorry state.'

HOLDING BACK THE TIDE

My visit to the Fens gave me a chance to call in at one of the most evocative English locations I've ever visited: Cley Marshes in north Norfolk. Though three years had passed since I last saw these marshy pools and dense reed beds, it felt like yesterday. Almost every stage of my life has been touched by my relationship with this coastline. From childhood memories of my mum taking us to feed the ducks at nearby Salthouse to spending weeks as a bird-crazed teenager sleeping in a shelter that became known as the 'Beach Hotel'. Then there were camping holidays in my twenties, followed by the twitching years in middle age as I chased storm-blown waifs and strays like the isabelline wheatear, the bird that had drawn me back on this occasion.

My visits to Cley are now much more occasional, but the pull of a rare wheatear was too much to resist. A dozen or more of us stood marvelling at how tame the bird was – all the better for capturing this avian gem on film.

Cley Marshes, a reserve owned by the Norfolk Wildlife Trust, has long been protected from the advances of the sea by a steep shingle seawall. I remember the bank being steep and narrow, but on this visit, I noticed how the defences had been greatly strengthened. Keeping the sea at bay is clearly becoming more of a challenge.

With sea-level rises from climate change likely, I was reminded of something Ian Rotherham had told me: the bigger the engineering solution, the greater the catastrophe when it breaks. I couldn't help but wonder how much longer the marshes will survive the ravages of the sea. How much of the marsh might get swept away if the Wildlife Trust can no longer hold back the tide? What would happen to the view, to the local church, to Cley Windmill and to the pretty collection of shops and houses that make up this most picturesque of coastal villages? If the wall falls, where will the people and wildlife go? These are the questions that are at the back of many minds in low-lying regions, but for the Fens and its farming communities, they are not new. Ever since the first attempts to drain the Fens a thousand years ago, a battle to protect its soils from flooding has been ongoing.

Some sixty miles of seawall protect the Fens today, and recent projects to strengthen key areas have seen seawalls built that are up to seven metres high. Even temporary flooding can render farmland useless for years due to the effect of salt on the soil.[21] Farming leaders warn that Fenland flood defences are inadequate. The National Farmers' Union has been calling for greater investment in flood defences to protect farmers and food manufacturing. The Environment Agency claims to be investing millions of pounds to solve an 'unprecedented, complex problem', while DEFRA has pledged £2.6 billion of funding over six years and £1 billion for flood defence maintenance.[22]

For how much longer can awe-inspiring feats of engineering hold back the tide and prevent hugely important areas like the breadbasket of the Fens from long-term flooding? As the world warms up, sea levels are expected to rise by at least metre by the end of the century. And as Ian Rotherham reminded me, mean sea-level rises don't account for tidal surges that can be three metres above the average high water mark in some places. 'The land is shrinking, the sea's going up, extreme events are more frequent and more extreme, and it seems to me common-sense that it's not sustainable', he said.

Rotherham is not the only voice issuing warnings about the future of breadbasket landscapes. A 2015 report on global food security agreed that the world's major breadbasket regions could be hard hit by climate change, issuing a chilling warning of 'simultaneous multiple breadbasket failures'. The report goes on to say that we can expect more shocks in future: What we would call a rare extreme food production shock in the late twentieth century is likely to become more common in future.'[23]

Given that the Fens and other low-lying arable areas are often breadbasket landscapes, crucial for so much of our food, I asked Rotherham for his view. 'We may only have thirty to fifty years left of that, and you're [then] entering a situation of global food insecurity and increasing human populations. This doesn't sound like a desperately good scenario', he said.

RESTORING THE GREAT FEN

With such a rich landscape facing existential threats from both land and sea, there is a growing search for new ideas and approaches to preserve the Fens for the future. One such approach is a major project to restore the Great Fen, which might set the scene for the return of the cranes. The idea is to reconnect the nature reserves of Woodwalton Fen and Holme Fen, creating a 3,700-hectare wetland with seasonal pasture. With 99 per cent of the wild fen having been destroyed by drainage, this ambitious new project would go a long way to preserving the unique Fenland ecosystem.[24]

On their own, these two nature reserves were too small and isolated to support the special Fenland wildlife that had survived.[25] 'As you can see, they're surrounded by very intensively farmed arable land', said Kate Carver, who spearheads the project for The Wildlife Trusts. 'There's only about 1 per cent left of natural fen … and these are two of those very precious remnants.'

The big danger for isolated reserves is that their effect as nature's 'time capsules' may start to diminish. As the surrounding countryside becomes more barren, it undermines the richness of the nature reserve by the slow asphyxiation of isolation. After all, nature rarely respects boundaries. I see this at my local nature reserve, Farlington Marshes in Hampshire, where declines in bird numbers are evident, even though the habitat itself remains undiminished. Birds move about, nesting in one place and feeding in another. They might spend the summer in one landscape and the winter somewhere else. For long-distance migrants, that means different countries, but for others, the need to switch between places can be much more localised. And because of this, they need joined-up landscapes to thrive.

The 'holy grail' is a combination of landscape-level initiatives like the Great Fen with nature-friendly farming. What I found compelling about the project was that rewetting a large area of Fenland could help a beleaguered farmland landscape by buffering against flooding, preserving peatland soil and reducing carbon emissions, all while encouraging wildlife.

A key to this venture's success lies in restoring farm animals to the land as ecological grazers. Four-fifths of the pasture on the Fens has been ploughed under,[26] and restoring grassland is an important part of the Great Fen initiative. It helps to stabilise the peat and stops it disappearing into atmospheric carbon. It also weans the land off artificial fertilisers, which would otherwise favour common arable 'weeds' that choke natural flora before it gets going.

A network of farmers and graziers now grows hay on more than five hundred hectares, while much of the rest is grazed by fantastic-looking animals like British White and Belted Galloway cattle that help to provide a diversity of habitat, leaving long clumps of grass in some places and grazing tightly in others. All this serves to provide habitat for ground-nesting birds and other wildlife.

Chris Wilkinson is a farmer whose Norfolk Horn sheep and Aberdeen Angus cattle graze three hundred hectares of the seasonal grassland that is re-emerging in the flat Fenland landscape previously dominated by intensive arable farming. His 140 breeding cattle and their calves are pasture-raised: 'We don't push them with concentrates – they're grass-fed', Wilkinson told me. 'The meat's more tender, so you get a better eating quality.'

Then there are the insect hotspots created by cowpats. 'There is a massive ecosystem that develops around a cowpat, which is why I don't use wormers.' Chemical parasite treatments make cow manure smelly and sterile, whereas natural pats provide great wildlife potential.

Wilkinson sees the loss of organic matter in arable soils as one of the big problems in today's agriculture. 'I'm very concerned because farming is being market-led down the path of intensification and specialisation, and with that comes difficulties with rotations and good agricultural practice... There is an issue with soil because of the fact that we're specialising and we're not putting any organic matter back.'

However, despite the perilous state of the nation's soils, Wilkinson sees signs of a reversal in this trend. 'There is a move now to reintroduce livestock. And it's happening in East Anglia, with farmers who are arable specialists reintroducing sheep and cattle.'

Bringing farm animals and arable production together needn't mean eating more meat; in fact, as Wilkinson told me, 'We don't need to eat as much as we do.' However, if we're going to farm for the future, it seems like restoring this 'contract' with the land, ending the segregation and rotating these major elements in harmony, is essential.

Fenland peat soil is among the most productive in the country, but it is also the most severely threatened. This very basis of Britain's breadbasket landscape is disappearing at the rate of two centimetres a year. And this gets to the nub of a criticism levelled at the idea of reflooding prime farmland: what if there is a situation where no patch of soil could be spared in producing food?

'If there was ever some national emergency where every square centimetre of land was necessary, you could drain all this and you could farm again', Kate Carver told me. 'Whereas if you carry on farming intensively, in thirty years' time, there's not going to be peat left in which to farm.' She also pointed out that 5,000 hectares of East Anglia is used to grow flowers rather than food.

To date, the Great Fen project is nearing the halfway mark, with 1,700 hectares under restoration. According to Carver, recreating the entire Great Fen could take fifty to a hundred years, especially given the reluctance of current landowners to sell prime arable land. However, my time in the Fens had taught me that, with the land shrinking and the sea rising, this particular breadbasket may not have another fifty years left.

2

A TALE OF TWO COWS

AROUND THE WORLD IN EIGHTY MINUTES

Outside our bedroom window is a cow barn. It is so close that I could throw a stone and hit it. Ash-grey corrugated roofing is freckled with yellow and green lichen. Grey breeze blocks inside divide its straw-lined interior from an open-air yard. The barn provides winter housing for the forty cows that graze the surrounding river valley pastures in spring and summer.

In 2020, I watched the cows released after winter on a mid-April day so warm that it felt like summer. Still young, they were literally twitching with excitement, their eyes wide and their ears up. They were steers, male animals, with athletic-looking frames in a range of colours ranging from black-and-white to orangey-brown. They were a mix of dairy and beef breeds. Some of them were gorging on lush grass, a new sensation for animals used to dry winter fodder. A few paused their grazing to reach the fence and meet noses with our dog, Duke. At that moment, the line between cow and dog became blurred as they greeted each other. They were bonding, offering mutual reassurance in a way that Gavin Maxwell noted of his otters in *Ring of Bright Water*. They performed the ritual that is important for so many animals: licking each other and exchanging saliva.

As I watched the herd in the sun, it was clear that each cow was different. It reminded me of the words of the cattle farmer Rosamund Young, who described them in *The Secret Life of Cows* as being 'as varied as people. They can be highly intelligent or slow to understand; friendly, considerate, aggressive, docile, inventive, dull, proud or shy.'

Cows have their own language, too. They moo in fear, disbelief, anger, hunger or distress.[1] Scientific research suggests that they 'talk' to one another and retain individual identity through their voices.[2]

As Duke and I stood with the herd, a strained moo echoed along the valley; a lone animal stranded on the other side of the river was expressing fear at missing out. The rest were clustered on lush grass and wildflowers beside the ancient bridge. Two were sparring, their heads pressed together in a bovine arm wrestle. Another rubbed against a telephone pole to relieve an itch.

This is far from factory farming, and much closer to the image used by marketeers of the countryside. Yet so much of what gets marketed like this shouldn't be. Things have been changing greatly since the Second World War. In Britain, nearly three-quarters of the land surface is farmland, two-thirds of which is pasture. Yet most farmed animals are indoors, caged, crammed and confined. Britain has a long tradition of keeping cattle and sheep on pasture, but for some years now, the reality for cattle has become less palatable.

Investigations published in the *Guardian* have revealed that Britain now has nearly a dozen US-style mega-farms rearing cattle for beef. Industrial-scale units, each big enough for up to 3,000 cattle, were found in Kent, Northamptonshire, Suffolk, Norfolk, Lincolnshire, Nottinghamshire and Derbyshire. Rather than being grazed or barn-reared, the cows were held in grassless pens for extended periods.[3]

Although still rare in Britain, intensive beef-rearing 'feedlots' are common in the US, where they are known as concentrated animal feeding operations (CAFOs). However, the revelations originally published by the *Guardian* that US-style feedlots had come to Britain contradicted assurances by DEFRA Secretary of State Michael Gove. He had previously told Parliament that

Britain's departure from the European Union would not result in the spread of such mega-farms: 'I do not want to see, and we will not have, US-style farming in this country.'[4] His well-intentioned words were coming too late.

The mismatch between Gove's welcome statement and reality illustrates the insidious creep of intensification. His words came after new research by Compassion in World Farming showed that the spread of mega-farms was not confined to cattle. Britain had nearly eight hundred mega-farms for pigs and poultry as well as cattle, each housing as many as a million chickens or 20,000 pigs.

While the number of intensive farms for beef remained small, the trend was nonetheless worrying. Data for 2018 suggested that there were at least 384 intensive beef farms in England and Wales where cattle were confined permanently with little or no grazing, and that could be the tip of the iceberg. Data is only collected for farm units that fall within areas affected by TB, and therefore operate under disease-control restrictions. For most cattle farms, therefore, information on how the cows are kept simply does not exist. And though the march of US-style mega-farms seems a million miles away from the forty cows that graze in the river valley near home, the question is: for how much longer?

GOOGLE EARTH

When I watched my local cows enjoying the spring grass, I should have been on a work trip to the US. I had planned on visiting some of the biggest cattle mega-farms in the world, but the Covid-19 lockdown had thwarted my travel plans.

Undaunted, I decided to conduct that travel through the Internet. For years, I have taken every opportunity to witness what goes on in factory farming. To see the sights, sounds and smells. To meet the characters. To discover what goes on behind closed doors or locked gates. And I found that by travelling virtually, I could, of course cover much more ground without leaving home.

And so it was that I started my 'Google Earth world tour' of some of the biggest cattle feedlots on the planet. I found them in the Midwest plains of Colorado, Nebraska, Oklahoma and northern Texas. Others were in Idaho and California.[5] My search for the 'world's biggest' led me to the J. R. Simplot Company near Boise in Idaho, in particular the company's Grand View feedlot, which holds 150,000 cattle[6] on a single 750-acre site.[7]

According to a candid assessment by Iowan farmer Michelle Miller, 'It's a real-life "corporate" farm that may also be referred to as a "factory farm".' An article in the industry journal *AgDaily* invited readers to 'Take a look inside one of the nation's largest cattle feedlots'.[8] I took my cue and entered the farm address into Google Earth. Before I knew it, I had an aerial view of the Midwest Plains. As I zoomed in on my destination, I felt queasy, as if heading down fast in a lift. I now had a drone's-eye view of an impressive office building with three white cars parked outside. Trees lined a lawn, while flags fluttered on flagpoles.

I was soon staring at the feedlots with tens of thousands of cattle. It was a flat, featureless expanse of barren pens with little or no shelter, and not a blade of grass in sight. It went on and on. I had seen drone photos of places like this before, with countless cows as specks on never-ending scorched earth. Now I was able to take identical pictures from the comfort of my own home. It felt surreal to be hovering above 150,000 cows nearly 5,000 miles away. Staring intently at my computer screen, I zoomed in closer and could make out individual cattle, standing 150 to a pen. According to Michelle Miller, a team of up to eighteen 'cowboys' patrols these pens each day on horseback. However, I couldn't see the need to ride horses here – after all, the cattle were rounded up permanently.

With more than 26,000 industrial cattle farms in the US, most of which hold over a thousand cattle, feedlots are big business there.[9] Instead of grass, animals are mainly fed cereals: corn, wheat and barley, along with soya and leftovers like distillers' grains.

Miller asks, 'Do we "feel" like farming should be romantic by Old MacDonald and a handful of animals?' Or should we get real and have intensive feedlots for an 'upcoming population of 9 billion to 10 billion people (and billions of pets to feed)'? Her question is one that is being debated by governments and civil society all over the world.

Courtesy of Google Earth, I travelled on to South Africa. A Bloomberg report suggested that a 160,000 cattle-capacity site at Karan Beef feedlot south of Johannesburg would trump anything I'd seen in the States for size. Glued to my laptop, I searched the area and zoomed in first on one housing estate and then another. But these settlements were dwarfed by the massive fan-shaped outline of the feedlot. Like the one I'd seen in Idaho, it was an expanse of bare earth divided into blocks, teeming with cattle.

Industry video footage gave me a ground-level view of pen after pen, all crowded with cattle, with no shade from the intense sun.[10] According to a commentary, the cattle are 'hormone-treated' on arrival. But although this farm fattens half a million animals for slaughter every year, most of the beef is destined for export markets in the Middle East and China.[11]

My next stop was Australia, where 450 feedlots worth $2.5 billion are said to provide 40 per cent of the country's domestic beef supply. Whyalla Beef's 56,000-cattle feedlot in Queensland has been described as the country's largest, with exports that reach Japan, France and Belgium.[12,13,14,15] Hormone growth promoters are 'commonly used' to boost weight gain and reduce the time it takes to rear an animal for slaughter.[16] And its sheer size means that cattle are sourced from across the country, sometimes as far south as Tasmania.[17] The thought of animals making such long journeys made me shudder. The whole thing seemed so out of kilter with the industry's ambition to supply 'clean, green, disease-free' beef.[18]

My whirlwind tour of the world's biggest industrial cattle farms left me strangely elated – in little more than an hour, I'd seen more mega-feedlots than I'd visited in my entire career. I hadn't felt the hot dust under my feet, gagged when the stench hit the back of my throat or looked the animals in the eyes, but I'd done that plenty of times before. Armed with know-how I'd gained from 'visiting' eye-catching examples of mega-farms, I soon found myself coaching newspaper journalists on how to 'tour' the world's mega-farms by Google Earth.

As I closed my laptop, our neighbour's forty cows were louder than I'd ever noticed. It made me wonder, had my Internet tour heightened my senses? Or was I now more aware of the herd, having just stared at hundreds of thousands of their confined kin? Well, no. By sheer coincidence, the cows were standing in our garden – all forty of them! As they were being moved from one pasture to another, they'd made a run for it. It was a reminder that cows have minds of their own and a desire to feel joy if we let them. As Rosamund Young would no doubt explain, they had imposed their own will on the situation. And for us, those forty cows had left their mark on our property: cowpats on the lawn!

KING FOR A DAY

It was early autumn and I had been invited to the New Forest home of one of Britain's best-loved TV naturalists, Chris Packham. His agent had given me a half-hour slot for the interview, so I shoehorned it into my diary before an important board meeting. Two hours later, I was still there – I had to hope my trustees would be understanding. Watched intently by his beloved miniature poodle Scratchy, Packham was warm and generous with his conversation, as we talked about our beginnings, the countryside, food and much more. He offered me coffee and I asked for it black. But as he opened the fridge, I spotted cartons of plant-based milks and changed my order: 'Organic soya, please.'

Packham was just back from his daily walk in the New Forest, and his enchantment with the natural world was obvious. He'd crouched on the ground to investigate a type of fungus that was new to him, feeling its texture and watching as gnats landed on it in search of somewhere to lay their eggs. He took a photo to help with identification. Even its smell fascinated him. It was all part of an ongoing journey of discovery, where the colours, shapes, textures and smells of the forest form an uplifting part of life. And living in the New Forest, what a canvas he had to explore! It remains one of southern England's largest unenclosed tracts of pasture-woodland, forest and heath. As well as fungi, it is home to ponies, cattle and even donkeys, all of whom graze freely and help to maintain a patchwork of different habitats. In a tradition dating back to William the Conqueror, pigs roam the woods in autumn, eating nuts that might otherwise be poisonous to ponies.[19] Five species of deer wander beneath canopies of oak and beech.[20] Grass snakes, adders and lizards bask among the heather. And birds abound: breeding species include Dartford warbler, honey buzzard and lesser spotted woodpecker.[21]

Packham has a deep-seated fascination with the natural world and a greater understanding than most. Whereas many kids grow up wanting a puppy or rabbit, he wanted a kestrel, a bat or an otter. I suspected that, like me, he was inspired by works like *Kes* and *Ring of Bright Water*. He was born and brought up near Southampton before moving to the New Forest. As he wrote in his moving memoir *Fingers in the Sparkle Jar*, it was there that he discovered that 'wasps drank, newts gulped, skaters skidded, everything was new. Everything needed knowing.'

Chris and I both grew up in the 1970s, an era when wildlife around the farmland that covers most of Britain was still pretty abundant. That said, there was no instant gratification. Finding things for yourself and getting close to wild creatures also meant spending time watching, waiting and hoping. Packham gained an early appreciation of the simple beauty of life from having a ladybird on a finger or tadpoles in the palm of his hand. 'It was those simple things which are undeniably and completely beautiful,' he told me.

His thirst for knowledge about the natural world is unquenchable. 'There's still so much to learn. There's still so much out there that needs knowing.'

Yet Packham is clear that we know enough about the countryside to stop doing the things that are ruining it. A glance at the headline data supports his view. Since the 1930s, Britain has lost 97 per cent of its wildflower meadows,[22] which means less habitat for essential pollinators like bees. Farmland butterflies in Britain have declined by 37 per cent since 2005 and 57 per cent since 1979.[23] Populations of farmland birds have, on average, more than halved since 1970,[24] with skylarks, starlings and lapwings among the species that have suffered serious declines. Lapwings now have the dubious reputation of being the most rapidly declining bird species in Europe.[25] And industrial agriculture is one of the most important drivers of nature's decline.

More is known about the state of nature in Britain than anywhere else on Earth. There are few countries where a greater proportion of land is farmed.[26] And how the nation's birds fare has been accepted by the UK government as an indicator of sustainability and our quality of life.[27]

The fact that species which have suffered serious long-term declines are still declining is particularly worrying. Despite efforts by conservationists to stem the flow, our 'natural capital' – the overall state of the ecosystem that sustains us – continues to ebb away. Four farmland bird species – corn bunting, grey partridge, turtle dove and tree sparrow – have declined by more than 90 per cent since 1970, and their numbers continue to dwindle. According to government data, grey partridges have declined by 19 per cent in the last five years, while turtle doves have slumped by a further 51 per cent.[28] Surely the next stop is extinction.

Packham argues that farming has been a healthy part of the environment for thousands of years. It was always diverse, opened up opportunities for wildlife and was largely in harmony with the ecosystem. There was a fusion between food, farming and nature. However, he now sees wildlife and the countryside as being in 'desperate trouble', and he is not alone. The government's annual assessment of wild bird populations cites the loss of mixed farming as one cause of decline. An increase in the use of chemical pesticides is also blamed, a consequence of the growth of monocultures, which lose their natural resistance to 'pests'.[29] When farm animals are moved from the fields into factories, everything suffers: animal welfare, wildlife, soil and the sustainability of food production.

'It's all about dominating nature. It's about controlling it and generally controlling it without tolerance. It's about pesticides, it's about herbicides. But what we're doing is damaging that environment beyond the point that it will support our farming processes. We know that soils are in a disastrous state, not just in the UK but all over the world. Soil is fundamental to any terrestrial farming. If we don't look after those soils, we're doomed.'

If Packham were king for a day, this war on nature would stop. However, vested interests present a barrier to saving the countryside. Small farms have struggled while large farms have prospered, often backed by multinational agrochemical and grain companies, whose lobbying can hold great sway. Perhaps the biggest effect these vested interests have is on dampening progress at policy level, muddying the waters enough to suffocate any gathering of political will to change.

However, Packham cautions against blaming farmers per se and wants to encourage farmers who are nature-friendly and support animal welfare. 'It's not all farmers', he said. 'Farmers are not bad people. Some of the things that some farmers do are very bad – there's a clear difference. There are certainly some bad farmers out there, but there are some very good farmers, too.' As he sees it, we have to support those farmers who understand the problems and are doing something about it, the ones who see sustainability as 'embracing a care for wildlife and the environment'.

Yet it can't all be left to individual farmers or consumers to stave off farmageddon. Deep-seated problems need big solutions, and that means policymakers. For far too long, politicians have tended to deal with more immediate matters that will help get them re-elected, rather than tackling some of the big challenges coming down the tracks. And the future of food and the countryside is one of the biggest.

Packham wants to encourage a unified movement, a broad-based alliance for environmental change. In 2018 he organised the 'People's Walk for Wildlife', a march through London of 10,000 people holding smartphones playing birdsong. The march culminated with children handing in a statement to 10 Downing Street calling for an end to the war on wildlife. It was an event to draw attention to the interconnected issues of the countryside, farm animal welfare, wildlife and food. As Packham put it, it was for people who 'care about life'. 'We've got to see the bigger picture in order to make a faster difference', he said.

Packham sees the problem not as a lack of knowledge, but a lack of will on the part of the politicians. After all, naturalists have been documenting declines and shining a light on the problems causing them for decades. We know what needs doing, so why isn't it being done? Packham says, 'I've learned that human beings don't like changing their mind. That's the problem. We come up with new science, we come up with new ideas, we come up with proven methods, and people still won't change their mind. You show them a better course and they still won't take it.'

He's frustrated that we have the answers to solve the problems, but we're just not doing it. 'I know full well, given the resources, I could start this afternoon. If I was king for a day … I could start this afternoon. I could make a difference for wildlife by this evening.'

Then why isn't it being done? He thinks part of the problem lies with decision-makers who don't know enough about the natural world to make the big calls that are going to count. 'I don't think we're yet in a position to elect decision-makers who are particularly educated in looking after biodiversity.'

Climate change is a good parallel. Two to three decades ago, most politicians weren't aware of global warming, but now it has become part of their job description. Though of course more needs to be done, at least the political awareness is there. With biodiversity and the plight of the countryside, that is not the case.

'These people, they don't understand fundamentals of ecology, some of them, I fear, haven't got past *The Ladybird Book of Ecology*, or in fact page 1 of it. And so my hope is that, at the moment, the triage that we're applying, the sticking plasters that we're putting on, as rapidly and effectively as we possibly can, will keep enough of nature alive, up until the point that we have new generations of decision-makers who do understand, and they will make the right decisions, because they know they need to.' In the meantime, Packham emphasises the importance of picking a fight with the right people, which means backing those who are doing things differently and 'farming the right way up'.

FARMING THE RIGHT WAY UP

'It was dead ... there were no worms, there wasn't an insect ... The soil was hammered', Simon Cutter recalled. When he first dug into the soil on his 500-acre farm near Ross-on-Wye, Herefordshire, what did he see? 'Nothing. It was just solid mass, no structure. It's a difficult soil anyway, but it was particularly dead.'

Cutter's farm is marginal land. Arable crops used to be grown here, but when he inherited it, it was a mess, with stones and weeds everywhere. Previous farmers had fought the land to grow crops, throwing pesticides and artificial fertilisers at it. In fact, the land was in such poor state when Cutter bought it in 2004, people thought he was mad to take it on.

Farming here had been the wrong way up. What did he do? He planted grass, restored the land to permanent pasture and brought back wildflowers, all of which added fertility to the soil. Insecticides and other 'nasty chemicals', as Cutter described them, were dispensed with and nature let back in.

At his aptly named Model Farm, Cutter now keeps cattle and sheep in a welfare- and nature-friendly way. The farm overlooks Symonds Yat, a wooded gorge known for the limestone outcrop that rises five hundred feet from the banks of the River Wye. Bones of a sabre-toothed tiger were found in nearby King Arthur's Cave,[30] but any bones discovered today are more likely to be those of pigeons consumed by nesting peregrine falcons.

Cutter farms organically, and his cattle and sheep are grain-free. Conventional agricultural teaching in recent decades has been that rearing cattle and sheep relies on feeding them grain, but this turns farming and natural biology on its head. After all, ruminants like cattle and sheep are adapted for eating grass and other plants; feeding them grain can upset their stomachs, causing cattle to suffer acidosis of the rumen. It's a way of doing things that has taken hold because there is profit for feed companies in selling grain to livestock farmers.

Instead, Cutter's three hundred Herefordshire beef cattle graze rich pastures. The calves suckle on their mothers, their reddish-brown bodies contrasting with white heads, breasts and bellies as they nibble hillsides covered in tall seeding grasses and ox-eye daisies. Their cowpats are rich in insects, which benefits local wildlife.

Cutter also breeds a flock of 350 Wiltshire Horn crossed with Welsh Mountain sheep. These are 'easycare' sheep, resistant to foot rot and worms, which blight other breeds. Greater horseshoe bats, rare in Britain due to the use of chemicals and the loss of grazing animals, have recently returned here.[31] In this area at least, farming seems to be the right way up.

I stood with Cutter in a splendid field of red clover. 'That's where our protein comes from – we don't need imported soya. If you feed red clover to pregnant ewes, you don't need to feed any corn to keep them healthy', he said.

Clover keeps breeding sheep healthy and also helps the soil, avoiding the need for nitrogen fertilisers. Little nodules in the roots of clover pull nitrogen from the air and fix it in the soil. 'Smart, whoever thought of that one, because it doesn't cost anything.'

Together with a retail business where much of Cutter's produce is sold, the farm is profitable without relying on government subsidies. And its input costs are low because he doesn't use grain feed or chemicals. Would he recommend this to others? 'All livestock producers should go grass-fed. They would do much better. Their costs would be significantly less.'

In 2019, Cutter was awarded Compassion in World Farming's inaugural Sustainable Food and Farming Award for turning marginal land into pasture and providing high-welfare conditions while also restoring soil and nature. As we walked around the farm, I found a suitable moment to hand over the engraved glass trophy, and we posed for photos among his herd of Herefords. Cutter gripped my hand with pride. 'It should be the cows you give the award to today, really, because they've done the work', he said.

3

SIXTY HARVESTS LEFT

WALKING ON THE MOON

Early morning and a tractor was ploughing the remains of the previous year's pumpkins and wheat into the soil. Steel blades flashed in the sun as they cut through the earth, leaving ruts eighteen inches deep. When each row was complete, the blades lifted and the tractor bounced, clanked and snorted as it did an about-turn before carrying on once more.

I watched this toing and froing as the tractor ploughed its lonely furrow, dust clouds spiralling behind, catching the sun and creating an aura over the juddering vehicle. Only, something was missing: there were no screeching gulls following the plough for an easy meal of worms. Because, as far as I could tell, there *were* no worms. The soil seemed lifeless, which is why the gulls weren't there.

A few young gulls did eventually show up, no doubt attracted in their naivety by the tractor turning the soil and promising easy pickings. About two dozen of them swooped down behind the plough and alighted in its wake, but they soon gave up, instead gathering forlornly on cellulose strips put down to protect newly planted crop saplings.

It made me wonder what would happen if I came back with a spade and dug for worms. To be honest, I didn't fancy my chances. Then I realised that the tractor was ploughing across a footpath, allowing me the right to inspect the newly turned soil more closely.

I stared down at earth that had been ploughed just moments before, but there was no sign of anything moving. There were no worms or insects desperate to get back into burrows, their worlds having been suddenly turned upside down. Soft, sandy soil was all there was, thrown into gentle ridges. I walked deliberately, peering at the ground as the gulls had done. I kicked at clods with my boots to see what was inside, but each one was coarse and easily broken. It was lifeless, without a worm to be seen. No wonder the gulls had given up – I could have been walking on the moon.

Yet there was more drama to come. Most days, I walk with Duke across these fields. That day, he bounded in and out of the newly ploughed ruts, his ears flapping and tongue wagging. To him, it was a great game. On previous visits, I'd watched a pair of woodlarks courting in these fields. Small brown birds with short tails, their colour is in the male's song. For days, the melodic yodelling cascade of the male's song fell over this ground in echoing, mesmeric fashion. How many times has that sound lifted my spirits? How many times have I thought to myself that their song trumps that of their much-vaunted cousin, the skylark?

The song of the skylark may be much written about, iconic even, but I can't help thinking it is more to do with them being familiar. Few writers, unless they are birders or naturalists, would recognise the woodlark. After all, what we don't recognise, we often don't know is there.

Skylarks are much less common these days, thanks to intensive farming, and woodlarks are rarer still, preferring heathland habitats. We have valuable fragments of heathland in our area and the extensive heaths of the New Forest are not far away. However, here on the farmland around home, woodlarks are seen and heard in spring and autumn, while the once familiar skylarks have become the local rarity.

I'd previously watched the male woodlark claiming his breeding territory, hanging over these fields in fluttering flight. He'd also sing from the oak treeline that serves as one of the few remaining divisions between 'fields', and from the birch tree beside 'New Pond', a man-made irrigation lagoon where dabchicks breed and

pond skaters skate. Woodlarks are ground nesters, and I'd seen the
pair on the ground near New Pond, hidden among rough grass.

Today, the field was silent, but for the low rumble of the
tractor and plough. The field and the grassy area beside the pond
had been ploughed right up to the raised bank, stopping six feet
short of the trunks of mature oaks and coming within inches of a
nearby footpath. In disbelief, I looked back at the area of tussocky
vegetation beside New Pond, where Duke and I had been walking
moments before. Now the plough was there, and the habitat was
gone. Woodlarks are scarce and formally protected – a Schedule
I species under the Wildlife and Countryside Act – making it an
offence to intentionally disturb an 'active' nest. I should have said
something, and next time I will.

However, the ploughman probably never even realised they were
there, and in any case, he most likely didn't know what a woodlark
was. I remember Tim May, a farmer I have come to know, talking
about how detached farmers have become from the fabric of the
countryside and the wildlife within it. They are cocooned within
their tractors, with little awareness of nature around them and little
knowledge of it either. 'The really sad thing about me as a farmer
is that I can't tell you anything about birds. As farmers, you're just
taught that if it's not a crop, you don't need to worry about it.
That connection with the environment just doesn't exist within the
farming community. I think it's shocking that I can walk into a
wood and not know what I'm hearing. I don't know many farmers
who would know much.'

May, who farms 2,500 acres of land overlooking Watership Down
in Hampshire, is a nature-friendly farmer who has restored farm
animals to their ecological niche as grazing animals, in rotation
with crops. Back home, looking over this wormless, sandy field,
I recalled May telling me that the reason he switched to sustainable
mixed farming from a chemical-soaked, all-crop method was
because his soil was tired.

Here, beside New Pond, I had previously seen the rain washing
soil downstream, towards the river. Gulleys would open up, leaving
gorges across the field where exposed soils had been washed away.

Having lived in this area for a few years, I could see the downsides of intensive farming for myself; although what I was seeing was nothing on the scale of industrial agriculture in other parts of the country or in the US, it was intensive nonetheless. I recalled the local landowner lamenting how his tenants farmed right up to the field margins, causing the soil to run into the river. Now I could see it for myself.

In the fields where the woodlarks had been, I can't recall ever seeing them in pasture; and I've never seen our neighbour's forty cows grazing there. In mixed farms, where crops are followed by animals, the pasture part of the rotation is the resting period for the soil, a chance for it to increase in fertility. This is also when carbon gets stored back in the soil.

Sadly, much farmland is subject to enforced separation, where animals and croplands never meet. The east of England is the arable breadbasket where crops are grown, while the west and the uplands are where animals still graze. And all over the country, many animals have been taken off the land completely, separated from the soil and put into factory farms. With worms, woodlarks and soils disappearing, and not a farm animal in sight, one can only wonder what this countryside will look like to future generations.

EARTH'S FINAL FRONTIER

Imagine waking up to the news that scientists had discovered a new world, teeming with life. A world barely explored, where the number of creatures could take decades to name and identify. Well, that is essentially the world beneath our feet. Soil is the Earth's final frontier, and each small handful holds more living organisms than the number of people on the planet. We rely on it for nearly all of our food, for a stable climate and to stop rainwater running back out to sea. Yet across the world, we treat soil like dirt.

That wafer-thin skin covering the Earth holds a quarter of the world's biodiversity.[1] Worms, fungi and creatures teem aplenty, at least in healthy soils. Around the world, there are some 30,000 worm species and 5 million species of fungi that help bind the soil and soak up water. Soil even has its own microbiome containing at least a million species of bacteria.[2]

We can only imagine how NASA scientists would react if their Mars rover probes turned up anything like as much biodiversity. As it was, in the mid-1990s, the merest hint of life on Mars sparked headlines like 'Mars lived, rock shows meteorite holds evidence of life on another world' and 'Fossil from the red planet may prove that we are not alone', all ignited by the most tentative evidence of fossilised bacteria from billions of years ago.

Yet the hidden world that we walk upon every day gets scant attention. Indeed, after 10,000 years of civilisation, much of the life forms that comprise our world remain unnamed. It was Leonardo da Vinci who said, 'We know more about the movement of celestial bodies than about the soil underfoot.' Five hundred years later, things haven't changed much.

What is even more remarkable is that this same hidden world holds the key to our own survival. Soil not only has a richness of life that would give any rainforest a run for its money, but it also stores carbon from the atmosphere, soaks up water and recycles nutrients, thereby holding our planetary life together.

That covering of topsoil across the world, with an average depth of about a foot, holds nearly twice as much carbon as the atmosphere. If we treat it well, it could hold even more, presenting a powerful piece in the jigsaw of how to solve the climate crisis.

The water stored in soil is the source for 90 per cent of the world's agricultural production and represents about two-thirds of all fresh water.[3] Soil holds that water close to thirsty plant roots, making for productive farmland. However, even moderately depleted soil holds less than half the amount of water of healthy soil. The rest runs off into rivers and streams and back out to sea, carrying nutrients and also chemicals that cause waterways to become polluted. In worst-case scenarios, this leads to vast areas of sea or inland water becoming so polluted that nothing lives, creating dead zones. The more organic matter – carbon – in the soil, the more water it can hold. Healthy soils can stop homes flooding and prevent crops from wilting during drought.

As well as water, healthy soil supplies the nutrients and oxygen that crops require to grow. It also supports growing roots and protects them from extremes in temperature.[4] If we treat it right, we could produce a lot more.

According to the UN Food and Agriculture Organization, with sustainable soil management we could produce up to 58 per cent more food.[5] Yet instead, we choose an industrial approach that turns soil into the muck that we plant seeds in and where we feed them with artificial fertilisers. Many of the farm animals that once lived on the land disappeared inside factory farms, and those that weren't were separated from crops and kept on farms specialising in beef, dairy, sheep, pigs or chickens. The concept of mixed farming became a thing of the past, as the chemical age of industrial, or 'intensive', agriculture was born.

Ploughing with heavy tractors and applying chemical pesticides to the land undermines the soil, disturbing ecosystems and releasing soil carbon into the atmosphere. Heavy vehicles compact the soil, so water flashes off rather than soaking in, making it harder for plant roots to find their way. In half a century, tractors have become six times heavier and compaction has become a big issue globally, affecting an area more than three times the size of the UK.[6]

Artificial fertilisers kill bacteria and fungi that would otherwise break down nutrients and make them more available to plants, damaging natural soil fertility.[7] The age-old nutrient cycle of rotating crops and replenishing soils with manure from free-ranging animals has all too often been abandoned. Long-held wisdom recognised the benefit of following soil-depleting crops like cereals with replenishing legumes that fix nitrogen in the soil. Of returning crop residues to the soil as green manure. Of resting tired soils under fields of grass to allow them to recover. Free-ranging animals would boost the regeneration through their manure, returning partially digested dead stuff to the living ecosystem beneath their feet.

The bottom line is that soil is a finite resource: if we carry on as we are, it will continue to degrade and disappear. Calculations suggest that soil is eroding up to a hundred times faster than it is being formed.[8] A third of the world's soils are already degraded,[9] but such assessments are thought to be underestimates, because 'intensive fertiliser application may be masking land degradation'.[10] And things are getting worse – according to a review in *Science*, about 1 per cent of global land area is degraded every year. It goes on to warn of a 'false sense of security' due to the 'unsustainably high use' of fertilisers and irrigation.[11]

All of this has a bearing on our ability to feed people. In 2015, the UN FAO warned that 'unless new approaches are adopted, the global amount of arable and productive land per person will in 2050 be only one-fourth of the level in 1960'.[12] Healthy soils are generally made up of 5 to 6 per cent organic matter, but half of Europe's soils have 2 per cent or less.[13] Much lower than that and yields could be affected.[14]

Soils are degenerating at an alarming rate; relearning the secret of keeping them healthy is imperative if we are to reverse the trend.

SIXTY YEARS OF SOIL

'Soil is life', announced an official report by Rothamsted Research Institute, one of the oldest agricultural research institutions in the world. It was here in Harpenden, Hertfordshire, that chemical fertilisers were first developed almost two centuries ago. Now, over three hundred of the brightest minds are focused on applying science to the challenge of feeding the world.[15]

Beyond a modern reception area lies a 330-hectare site that includes a working farm and manor house. A marble bust of agricultural scientist Sir John Bennet Lawes, who founded the institute in 1843,[16] adorns the entrance. It was his invention of 'superphosphate' fertiliser that laid the groundwork for the widespread use of chemicals in agriculture.

I had come to the birthplace of artificial fertilisers to meet Professor John Crawford, the scientist behind the calculation that, on our current trajectory, soils could have only sixty harvests left. The irony wasn't lost on me.

For someone carrying the weight of such a heavy message, Crawford was relaxed and engaging. Leading the institute's search for 'integrated solutions' in agriculture, he was no soil geek. The Glasgow-born 55-year-old has a research career spanning cancer and Alzheimer's disease as well as ecology.

He had once seen his future in the stars rather than the soil. During an internship at the Anglo-Australian Observatory in Sydney, he made the startling discovery of the existence of a cloud structure on Venus.[17] The revelation made front-page news and launched a space mission, but despite his discovery, it was the call of the wild that guided Crawford's career. He had always been fascinated by the natural world. He learned about the perils of pesticides by reading Rachel Carson's pioneering book *Silent Spring* and was horrified by the Club of Rome prediction that by 2100, economic and population growth could outstrip the Earth's ability to deliver.[18] Influenced by the conservation icons David Attenborough and Jane Goodall, he started looking for a way to have a positive impact.

After a spell at the Scottish Crop Research Institute in Dundee, Crawford established a research centre on sustainable agriculture at the University of Sydney, which set the stage for his groundbreaking statement on soil. It was at a 'carbon farming' conference in Australia in 2010 that he first warned the world that soil could run out. In 2012, he expanded on the argument in an article for the World Economic Forum, writing, 'A rough calculation of current rates of soil degradation suggests we have about sixty years of topsoil left.'[19]

By 2014, Crawford's message was being amplified by the United Nations. At a speech to mark World Soil Day, the UN FAO's deputy director general Maria-Helena Semedo noted that if current rates of degradation continued, the world's topsoil could be gone within sixty years. She made no bones about why this was important. 'Soils are the basis of life ... 95 per cent of our food comes from the soil.'[20] Crawford's calculation had gone global; but how did it come about?

As Crawford explained, 'If you don't have all the information about something, you do a back-of-the-envelope calculation to know whether this is a problem you need to work on.' For him, the question was when the world's soils would run out if we carried on as we are.

Crawford and a colleague were talking about media interest in the environmental costs of meat production. He recalled how there was a lot going around about a hamburger costing a kilogram of soil. That got the two scientists thinking: how many hamburgers would you need to eat before the soil was gone? Crawford took a global estimate for the amount of topsoil in the world, the rate of depletion and how quickly it gets replenished. The bottom line is soil forms very slowly and you can lose it very quickly.

'When I did all that, it turned out that the account would run dry in about sixty years ... I thought, "OK, it's probably worth doing this a bit more carefully."' He was nervous of the backlash, admitting that there were people far better qualified than him to do the work.

However, he was clear that whether soils might run out completely wasn't really the point: it was more that soils are becoming scarce in a world with more people to feed. A reduction in the amount and fertility of soil would bring grave consequences long before it disappeared entirely. When soil becomes seriously degraded it becomes uneconomic to farm, meaning more farmland has to be found elsewhere. Since 1975, scientists estimate that nearly a third of the world's arable land has been lost to soil erosion.[21]

'And, actually the last thing we want to do is destroy more habitat by moving from one patch of soil to another', Crawford said.

As farmland moves on, wildlands are jeopardised and forests are cut down to make way for new fields.[22] This isn't just a theoretical possibility; at the current rate of loss, some 12 million hectares of agricultural land per year is rendered useless, an area equivalent to the arable land of Germany, Poland or Ethiopia.[23]

Although it was rough and ready, Crawford's calculation at the Sydney Conference in 2010 had started a conversation. In the wake of his icebreaker, other expert commentators made their own assessments. David R. Montgomery, professor of geomorphology at the University of Washington, estimated in 2012 that the rate of world soil erosion 'now exceeds new soil production by as much as

23 billion tons per year, an annual loss of not quite 1 per cent of the world's agricultural soil inventory. At this pace, the world would literally run out of topsoil in little more than a century.'[24]

A year later, scientists David Pimentel and Michael Burgess wrote that 'Each year about 10 million hectares of cropland are lost due to soil erosion, thus reducing the cropland available for world food production ... Overall, soil is being lost from agricultural areas ten to forty times faster than the rate of soil formation imperilling humanity's food security.'[25]

Suddenly, soil was seen as more than just 'dirt', but rather a matter of life or death for future generations. An inch of topsoil takes at least a hundred years to form, depending on climate and vegetation,[26] which makes it a non-renewable resource. Recent scientific assessment by the Intergovernmental Panel on Climate Change puts soil erosion from agricultural fields globally to be ten to one hundred times higher than the rate of soil formation.[27] In other words, we're withdrawing from the soil bank account much faster than it's being topped up. Before I met Crawford, I ran my own calculations based on UN estimates of farmland and widely published figures on the rate of soil loss. Topsoil depth varies, from thirty centimetres in deep soils to as little as one centimetre in thin soils.[28] My most optimistic result suggested that a fifth of farmland would remain in sixty years' time; at worst, all but 5 per cent would be gone.

Crawford had got people talking about soil as something that was valuable and running out, which has since led to other experts making big statements on the world's biggest stages: for example, in 2015 a group of scientists from the University of Sheffield told a UN conference on climate change that soil loss was an 'unfolding global disaster that will have catastrophic effects on world food production'.[29]

In 2013, Crawford returned to Britain and joined Rothamsted Research. 'Since coming here, I've really become convinced that there's a genuine emergency, that we need solutions quickly and at scale,' he said. 'Soil stores more carbon than plants and atmosphere combined; therefore, it is probably the single most important regulator of global climate after the oceans.'

Industrial agriculture relies on artificial fertilisers to feed crop plants but it doesn't feed soil microbes, meaning that they stop storing carbon and release it instead.

This begs the question, if we can feed plants with artificial fertiliser, why do we need the soil? 'Because the soil provides the plant with water and the climate with a sink of carbon,' Crawford answered. And if we keep feeding the plant and not the soil? 'We'll run out of water, we'll run out of food and we'll exacerbate climate change. Plants will suffer, too. As soil declines, the plants do less well; the soil holds less water so needs more irrigation, nutrients run off faster, causing inefficiencies and more water pollution: a classic case of the law of diminishing returns. And in a world of more mouths to feed and less resources, this isn't a way we can afford to go.'

SILENT DECLINE

Before soil disappears completely, it goes into decline. Some 40 per cent of agricultural soil around the world is classed as either degraded or seriously degraded.[30]

Crawford is clear on the urgency of action on this issue. 'The recent IPCC report made it very clear that land degradation is completely unsustainable, and we need to do something about it. And so we need to get into the business of soil regeneration.'

Industrial agriculture uses chemical fertilisers to mask soil decline, but an assessment by scientists at the Grantham Centre for Sustainable Futures described intensive agriculture as 'unsustainable'. Crop yields are maintained artificially through the 'heavy use of fertilisers', the production of which, they say, consumes 5 per cent of the world's natural gas production and 2 per cent of the world's annual energy supply.[31] Whichever way we look at it, declining soils will make a sustainable future a whole lot less likely.

Soil declines are an unintended consequence of breaking the bond between farmers, farm animals and the soil that has existed since the early days of civilisation. Intensive farming is short-termism at its worst: it aims to achieve bigger yields today using all the means that wreck the soil, meaning lower yields in the future. 'That's the issue that continues today', Crawford says. 'We still have this focus on yield ... but we're not thinking about the soil. If we didn't have fertilisers, we would notice our soil was degrading because our yield would go down. So we're compensating for crap soil by adding nutrients. Artificial fertilisers are fantastically successful in terms of feeding people. But if they are used in a way that doesn't feed the soil as well, more comes out of the soil and less goes in. And it will degrade over time.'

I was surprised to find support for Crawford's view from Andy Beadle, a spokesperson on sustainability for BASF, the largest chemical company in the world. Beadle believes that everything in farming begins and ends with soil, and that its importance is being realised only in the wake of plateauing crop yields. As he told me, 'Yield increases have completely plateaued out, despite all of the best scientific minds working on it. And to me, the root of that is soil. Because if we look at the soils now across Europe, the vast majority of them are in quite a perilous state.'

Companies like BASF spend many millions of pounds every year developing more productive varieties of crop plants. The industry has been spectacularly successful at driving up yields over recent decades, but the magic has stopped working. Beadle agrees with Crawford that there may be as few as sixty harvests left in the world's soils and points to the move from mixed farming to specialisation as a reason for soil declines.

'If you're an arable farmer in East Anglia now, you're not going to have any manure that you can put back on the land – partially composted farmyard manure is probably one of the best ways of getting carbon back into the soil.'

Crawford is certain that regenerative farming is the answer. 'In a way, we've known forever how to do it,' he told me. 'We just returned manure to soil – so there was always organic matter going into soil and we knew how to look after it.' His warnings have caught the attention of those at the highest level: he was invited to the White House when, during his last days as president, Barack

Obama was looking to rush through a strategy on healthy soil. The day-long workshop was attended by fifty delegates, including scientific experts and people from Hollywood. Obama could see that this serious message needed to permeate into wider culture.

'At the end of the day, it's not politicians or businesses, it'll be ordinary people that will change the world, and the way to enable that is to help them think differently and think more', Crawford reflected. And if we could only do one thing, I asked him, for the health of the planet? 'I'd say fix soil.'

PART TWO

AUTUMN

After the long, hot summer days, autumn had swept in with a vengeance, scattering the roads with leaves from the trees that line them. Driving rain had come in with a cold front; hundreds of house martins clustered in tight huddles around our farm, clinging to stone walls, slate roofs and every other surface that offered shelter. When the sky dried up, they were back on the wing, their dark-and-white bodies recalling tiny killer whales as they twisted and turned over the meadow, beginning their southward journey to sunnier climes.

As I walked with Duke along field boundaries, spinneys and riverside banks, there was an even greater sense of comings and goings; swallows, warblers and other birds were leaving for the south, while a dozen redwings, Scandinavian thrushes fleeing their northern breeding grounds, were a sure sign that winter wasn't far behind.

The white cotton tails of roe deer disappeared at the merest whiff of our presence, bounding away on springy legs and into cover. Shrivelled white mayweed flowers waved their wispy baubles in the breeze, while the herd of cows continued to graze along the river valley. As long as there's not too many of them, grazing can be a nature-friendly and efficient use of resources. In mid-November, as the ground becomes waterlogged, they'll return to the barn for winter

As the year progressed, the maize fields turned from green to acrid brown. Giant stems looked gaunt, their heads bowed as if in prayer.

Some of the corn still had cobs; these were either late developers or those missed by the migrant workers who toiled the fields.

Autumn rain showers poured off the maize-studded soil, drawing attention to obvious erosion. Thankfully, the grazing meadow prevented the river from bearing the brunt of soil on the move. When soil trickles into the river, it becomes silt and clogs up riverbeds. That change of name, however, masks the source of the problem: soil from industrial farmland moving to where it shouldn't be.

Returning home, Duke and I stopped by the cow barn. A brown sparrowhawk veered behind us, hopping over the hedge and hunting low, leaving a furtive party of long-tailed tits twittering in panic. The white outline of a little egret flapped along the river, illustrating how the countryside has changed in recent decades as a result of climate change. In the 1980s, British twitchers – rare bird enthusiasts – would travel miles to see these small herons, but now they are a regular sight from my window. The same goes for the Mediterranean gulls that have become regular around here – special birds from another place becoming commonplace.

As the countryside changes and the limits of our planetary resources become more obvious, so the impact of our lifestyle starts to bite, and the leaves of optimism crumple and fall.

4

THE MARCH OF THE MEGA-FARM

PIG SICK IN SPAIN

Huesca, northern Spain, and a giant mechanical upside-down umbrella is opened beneath an olive tree. Fruit tumbles into the waiting net as the tree is shaken, while a man on a tractor stands expectantly. This age-old harvesting process here has long been part of the rhythm of life.

The area is overlooked by the limestone massif of Sierra de Guara and the lush valleys and rivers flowing from it. The local village hall in Loporzano, the centre of this community for centuries, stands beside an ancient olive mill. Flanked by thick-trunked olive trees, some five hundred years old, the walls inside are lined with images of each of the fifteen villages that make up the municipality of Loporzano.

Sustainable tourism has revitalised this agricultural region. Mountain gorges are a magnet for adventure-seekers, rock-climbers, cavers and other tourists attracted by the fresh air and sunshine. A rich abundance of wildlife, from vultures to turkey-like bustards, has long drawn naturalists to the area. Yet this sparsely populated beauty spot is under threat of invasion: from factory farms.

'We're extremely worried ... if these farms are built, the impact on our lives will be huge,' said Rosa Díez Tagarro, a resident of Loporzano in the province of Huesca. In Huesca alone, there are 3.8 million pigs, compared to just 2.6 million in the whole of

Andalusia. Between 2012 and 2017, the Aragonese pig population grew by 6,000 a week.[1]

Born on the Spanish plains, Díez Tagarro moved from Barcelona to Loporzano with her husband Jaime. Together they renovated a ruined house in the mountains, from where she was able to work as a freelance translator. The couple had been drawn to this place by its peace and tranquillity, and they were not alone – people from across Spain, Britain, Italy and the Netherlands have settled here because of the quiet mountains, clean air and pristine rivers.

But they were in for a shock: a proposal had been lodged to build a factory farm near their home. 'I nearly fainted. We thought we were living in an area that was protected and that something like this couldn't happen', Díez Tagarro said.

Loporzano lies just south of the Sierra y Cañones de Guara Natural Park, a stunning limestone mountain range of deep canyons, gorges and caves, and much of the area is protected. However, the villages are excluded from that protection, and the factory farmers have moved to exploit this.

Industrial farming companies established themselves in the region by obtaining initial licences for modest new builds before scaling up. The original proposal in Loporzano was for 1,999 pigs, but villagers feared it could end up being a mega-farm with as many as 14,000 pigs. Then a second proposal was submitted for a similar operation that would see pig manure spread not far from the region's important archaeological sites and within sixty metres of people's homes. Villagers were up in arms, fearing the smell and the impact the farms would have on local rivers, wildlife and their way of life.

'At first, there was going to be one pig farm, then two. Now it looks like there will be more. It will be cheaper for them if they can have many pig farms in the same area', Díez Tagarro said. 'When I heard that I would have a factory farm from my window, that it would pollute local rivers and affect the birds, I decided to oppose it.'

Díez Tagarro called a meeting and made her neighbours aware of the plans. Villagers there had never protested against anything before, but the first meeting drew more than a hundred locals. A new organisation, Plataforma Loporzano sin Ganadería Intensiva ('Loporzano without Factory Farming'), established a dialogue with city hall officials. They set up meetings, enlisted public support and got their message into the media.

The organisation quickly became a symbol of a Spanish movement to save the countryside. They connected with other communities who had set up similar protest groups and encouraged others to do so. In September 2017, an umbrella organisation, the Coordinadora Estatal Stop Ganadería Industrial ('State Platform to Stop Factory Farming'), was established, bringing together groups from all over Spain.

I took part in Coordinadora Estatal Stop Ganadería Industrial's first meeting in Loporzano before speaking alongside Díez Tagarro and local dignitaries at a meeting in the Círculo Oscense, an imposing century-old building in the centre of Huesca. When I arrived there an hour before the meeting, I found a few elderly gentlemen huddled around, drinking and watching bullfighting on a big screen. It didn't fill me with confidence about the likely receptiveness of the audience, but how wrong I was. When the meeting opened, the bullfighting had been switched off, the lights were up and the place was packed. Locals were eager to join the battle to save the countryside. The discussion was lively and the room united. These pig farms had to be opposed and feelings were running high. The rural population was at war against an enemy that had established a foothold and now threatened to overwhelm.

I got my first taste of what the locals were up against when they took me to see a pig factory farm, beside a juniper tree. Approved with the agreement of a former mayor, the air was filled with the acrid smell of slurry – liquid manure. Only the smell and the odd muffled grunt gave a clue as to who or what was inside. A grey pickup truck passed by and a sudden feeling of not being welcome washed over us. Protesters had previously had pig slurry spread over their intended meeting site, but thankfully we didn't face such intimidation.

Factory farms keep pigs in barren, crowded pens, the mother pigs usually confined so tightly that they cannot so much as turn around for weeks at a time. The mind-numbingly boring conditions lead to frustration and suffering. The piglets mostly have their tails cut off, without pain relief, to stop them biting each other's tails. Tail docking is still routinely carried out on the majority of pigs, even though EU law forbids it on animal welfare grounds. The same miserable conditions that cause animal suffering are conducive to the spread of disease, and some 84 per cent of all antibiotics sold in Spain are used on farm animals. This is a big problem because the overuse of antibiotics in factory farming leads to the rise of antibiotic-resistant superbugs and the demise of these essential medicines in human patients.

The smell of the slurry 'makes you gag and gives you headaches', said Díez Tagarro. And with the influx of so many pigs, there's a lot of it. 'All it's going to do is pollute our rivers. We don't want those rivers to get polluted. They are used by local villages for drinking water.'

Spain's human population has now been surpassed by the number of pigs slaughtered annually. According to figures released by the country's environment ministry, Spain slaughters 50 million pigs a year.[2] And as pig numbers increase, so too do environmental concerns about an industry that produces more than 4 million tonnes of pigmeat and generates €6 billion (£5.4 billion) a year.

More than half of the pigs in Spain are farmed in Aragon and neighbouring Catalonia, and the animals' manure is usually stored in 'slurry lagoons', a major risk of pollution. The country's pigs produce enough manure each year to fill a stadium the size of FC Barcelona's Camp Nou more than twenty-three times over.[3]

Catalan authorities have reported frequent intentional dumping of manure into rivers.[4] Díez Tagarro told me that 42 per cent of the spring water in Catalonia had already become polluted. In a search for new sites for factory farms, Catalan agribusinesses have been forced to look elsewhere. 'Pollution is costing the Catalan government 6 million euros a year to deal with it. It's the taxpayers paying for it, not the factory farms,' she said.

In Loporzano, pollution and disturbance were already taking their toll on local wildlife. Bird species including the little bustard, black-bellied sandgrouse, calandra lark and Egyptian vulture used to be found here in decent numbers, and hundreds of red kites used to roost together. But not any more.

'We can still find them, but in very small numbers', said 59-year-old Josele Jsaiz Boletes, who moved from Barcelona to Loporzano over twenty years ago to set up a business geared around the area's wildlife.

Also lamenting the loss of birdlife is Miguel Angel Bueno. Known as the 'Vulture Shepherd', Bueno can be found surrounded by flocks of vultures. Trusted by the birds, he has for years been throwing them food from a wheelbarrow; in response, they gather round and some even land on his head.[5]

While the new wave of intensification may not affect the vultures directly, pig farms and the use of pesticides on surrounding crops will have serious consequences for the environment. 'Vultures may not be affected, but the insects will disappear, and without the insects, the smaller birds will disappear', Bueno told me. I asked him how he feels about the industrial farms themselves. 'I would like to dismantle the pig farms; open the pens and let the animals roam free.'

BIG AG IN BRITAIN

Nocton in Lincolnshire, and unexpectedly I found myself having returned to the site of a battle against the relentless march of the US-style mega-farm. A decade ago, this pretty fenland village became the epicentre of a row over whether cows belonged in fields or in mega-dairies. Out on the fen, rolling dark cloud stretched over a bulging sky. The land was flat and the fields sat two metres below the road, with a drainage dyke lower still. The fields seemed to stretch away endlessly. The parcels of land were barely distinguishable but for the vague difference in crop residues. Otherwise, there were no field boundaries and no hedges or bushes – the feeling of space, of being small in a big landscape, was overwhelming.

Nocton became infamous for a monumental environmental battle to prevent a proposed US-style mega-dairy. The plan was to keep 8,000 cows – the equivalent of nearly a hundred average-sized British dairy farms – indoors, where they would be fed grain instead of grass. Styled on the large-scale dairies of California, Indiana and Wisconsin, it was to be the first of its kind in Britain.

It was a dispute that galvanised local people, politicians, children, countryside campaigners, free-range dairy farmers, animal welfarists and environmentalists in nationwide opposition. There was a clash of ideology between those who believed that cows belong in fields and those who foresaw a future in factory farms. The man behind the proposal suggested in a radio interview that 'cows do not belong in fields' – words that came back to haunt him. Local schoolchildren drew pictures in protest and buses carried banners proclaiming 'cows belong in fields'. The row went all the way to Westminster, and more than 170 MPs signed a Commons motion opposing the 'super dairy'.

The plan was finally pulled when the Environment Agency stepped in over the pollution risk.[6] One dairy cow produces as much effluent as fifty people; a mega-dairy of 8,000 cows would produce as much waste as a city the size of Bristol.[7]

While villagers in this Domesday Book settlement saw off that particular threat, the march of industrial animal farming continues across the English countryside. The number of large industrial-sized pig and chicken farms in the UK continues to rise, with recent figures suggesting that nearly 2,000 can now be found across the country. Despite Michael Gove's assurance in 2017 that the country would not see US-style farming, the number of industrial-sized pig and poultry units in the UK increased by 7 per cent over the next three years, from 1,669 to 1,786.[8] Pig farms are classed as 'intensive' if they raise at least 2,000 pigs for meat or have 750 breeding sows. Chicken farms reach this classification if they house at least 40,000 poultry.

Lincolnshire has the fourth-highest number of indoor-reared livestock in the UK, with over 12 million animals confined inside. The counties ahead of it in the factory-farming stakes are Herefordshire, Shropshire and Norfolk.[9]

The march of US-style mega-farms – a step up again in terms of size – in the UK was first revealed in a 2017 investigation by the *Guardian*. Most of these farms had gone unnoticed, despite their size and the controversy surrounding them, partly because many farmers had expanded existing facilities rather than having developed new sites.[10]

New research for Compassion in World Farming revealed that by 2021, the number of US-style mega-farms in the UK had risen to nearly 1,100. This included 745 poultry mega-farms in England alone, with another 59 in Wales. The UK total is almost certainly an underestimate as official data for pig and poultry units in Scotland was unavailable due to cyberattack.[11]

In five years, the overall number of US-style mega-farms has increased markedly, a worrying trend that calls into question the country's claim to be a nation of animal lovers.

Industrial-scale farms have attracted attention because of local residents' concerns over smell, noise and the potential for pollution or disease outbreaks. Animal welfare campaigners argue that factory-style farming prevents animals from expressing their natural behaviour. There is also concern that mega-farms push smaller farmers out of business, leading to the takeover of the countryside by large agribusinesses and the loss of family-run units.

The battle against the Nocton mega-dairy proposal may have been won, but the attitude that went into the proposal prevails. To the industrialists, separating animals from crops, and farming both on a lowest-cost commodity basis, is the route to greater efficiency. The mass production of cheap ingredients is presented as the 'modern' way, and farmers are lured into becoming commodity-captured, reliant on subsidies and selling their wares for the lowest possible price.

CHINA CRISIS

I was taking the night train to Nanyang in China's Henan province in 2011 when I very nearly found myself alone and without possessions in an unfamiliar country. It was 2 a.m. and I was dazed with sleep. Wanting to stretch my legs, I stepped off the stationary train for a moment and looked bleary-eyed at the massive station. It was like something out of a James Bond movie, echoing and gleaming. For a moment, I was lost in thought. Then an alarm sounded and I jumped back on the train just in time, before the automatic door slammed shut and we pulled away. I felt shaken. It had been so close to speeding off into the night without me, leaving me to reflect on how well and truly stuck I would have been. However, less than twenty-four hours later I would find myself in a real jam, without my passport and at serious risk of not getting it back.

I spent the last hour before I reached my destination looking at mile upon mile of maize, which was largely grown for animal feed and biofuel. The air was dampened by a grey, misty haze. I was in China to witness how Western livestock breeds, techniques and equipment were powering a surge in the nation's agricultural industrialisation. I was keen to visit Muyuan, a company that was listed among the world's top ten biggest producers of meat and animal feed. They were reported to be rearing more pigs per year than any other company in Asia,[12] their production having grown from nothing in 1992 to a breeding herd of about 1.3 million sows.[13] To put that into perspective, the breeding herd of that single company was three times the size of the entire British pig industry.[14]

I had hoped to visit the company's headquarters but was refused an appointment, so I headed there anyway. I listened to the concerns of a group of about twenty villagers nearby, who as a result of pollution and mosquitoes since the farm was established, were unable to have their windows open during the summer.

Someone who overheard our conversation clearly had the interests of the company at heart: within hours, the police had seized my passport and those of my travel companions. At our hotel, we had been joined by a Mr Chan from Muyuan, who was keen to 'bump' into us. There was nothing for it but to slope off nervously and have some dinner.

Looking back, I think we were saved by the quick thinking of Jeff Zhou, our fixer and soon to be Compassion in World Farming's country director in China. Jeff burst into the room, accompanied by Chan and two companions, and exclaimed, 'Great news! We have new friends for dinner!' I pulled out a chair for Chan. He eyed us up, clearly regarding us as a threat, but I was keen for him to see otherwise. I reeled off all the companies we work with, and it seemed to work. We ordered dinner and continued talking. Before the end of the meal, Chan disappeared. So too did the police. When he returned, Chan invited us to meet Muyuan's managerial team the next morning. It felt like we had turned defeat into victory. And the best thing was that not only did we get our passports back, but Chan paid for dinner.

The next day we went on a tour of the company's newest 'farm', a site that resembled a housing estate with a couple of hundred low-rise buildings with concrete floors. They would confine tens of thousands of pigs. In the company's early days, Western farm equipment manufacturers dominated the market, selling their wares to companies like Muyuan that were new to factory farming. And it wasn't only hardware they sold. Muyuan uses a technique of 'super-early weaning' developed by a British company, where piglets are taken from their mothers at less than a fortnight old. Left to nature, piglets wouldn't wean before they are three months old. The company's original breeding stock were flown in from the US, the UK and Canada, and pregnant pigs were kept for prolonged periods in sow stalls, narrow metal crates where they can't turn around, a system long banned in Britain and the EU.

To be fair to Muyuan, they have since made some improvements to their animal welfare credentials, making commitments not to use sow stalls for pregnant pigs or painful mutilations such as tail docking and teeth clipping in some of its operations.[15] I had been proud to welcome the company's representatives to the stage on several occasions, to present them with awards recognising the progress they had made on improving animal welfare. And at each of these ceremonies, the Chinese recipients had always been among the most enthusiastic, which I always thought boded well for the future. However, my optimism has since turned to despair.

A decade on from my visit to Muyuan, I was staggered to learn that they were leading the way in developing multi-storey factory farms. Back when I first visited, they had twenty-one farms; they now have more than two hundred.

In a development that looks set to take mega-farming to a whole new level, the company is now trying to raise more pigs on a single site than anyone else in the world. Roughly ten times the size of a typical American breeding facility, Muyuan's new mega-farm aims to produce more than 2 million pigs a year. High-rise facilities are increasingly popular in China, thanks to a scarcity of suitable land.[16] Yet the disease implications – not to mention the animal welfare concerns – of keeping so many pigs in such a confined space make me shudder. This is ironic, given that it was a disease, African swine fever, wiping out around half the production of the Chinese pig industry that provided the initial impetus for these massive pig factories.

As well as showing me a hugely worrying new frontier in global mega-farming, my time with Muyuan highlighted the role of Western equipment manufacturers. Some of the biggest companies are based in Europe and continue to sell cages and crates globally, regardless of whether they are now illegal back home.

One of the leading companies in the field is German-based Big Dutchman, which claims to be the world's leading housing and equipment supplier for 'modern' pig and poultry production, selling to a hundred countries, including China.[17]

A quick look at the company's website revealed that it sells sow stalls.[18] Although they are widely used in China and the US, these stalls, also known as gestation crates, are banned for prolonged use in the EU and will be banned completely in Germany by 2030. Yet the company's sales literature describes them as a 'good method which allows the farmer to monitor and control each sow individually'. It continues, 'No matter if a farm is to be equipped in southwest China or in icy Siberia – Big Dutchman always offers the perfect solution to every possible problem.'[19]

Looking further into the company's sales material, I clicked on a video that showed five young chickens in a wire cage not much bigger than a microwave. Their beaks had been cut to ugly stumps – whoever kept them realised that their frustration might lead them to injure each other. The video cut to a shot of an endless row of cages. The confinement was claustrophobic – I could have been watching investigative footage from an animal activist group, but I wasn't. The promo video was selling 'conventional' cages for laying hens,[20] or *barren* battery cages as I tend to call them, a system long-since banned by Germany, the UK and the EU.

The standard rationale I hear for the sale of such industrial farming equipment is the importance of feeding the world, a self-serving excuse given the huge amounts of food wasted as animal feed for factory farmed animals. Government advisers and agricultural colleges in developing countries in Africa, for example, often promote cages and crates as the 'modern way'. Because they come from Europe, they have a cachet and a promise of modernity, yet not everyone in Africa welcomes this Western intervention.

'Cage farming is not a modern way of farming; it is a cruel way of farming', says Wachira Benson Kariuki, a lawyer from Nairobi who works in Kenya with the African Network for Animal Welfare. He sees the arrival of cages in Africa as a backward step and proposes export bans as the way forward in order to prevent Europe exporting its mistakes. As well as banning cruel systems, he argues that EU law needs to ban the means to perpetuate that cruelty elsewhere. Otherwise, 'they will come here and do the same thing they were doing in Europe'.

This is the perfect illustration of how industrial agriculture as a model has become locked in: legislation may tighten things up in one jurisdiction, but it continues to allow it to be exported, which spreads the problem further afield.

FALLING OUT OF LOVE

There are clear signs that the political establishment is falling out of love with the old way of intensive farming. When he was DEFRA Secretary of State, Michael Gove set about 'greening' agricultural policy in a way that would have previously been unthinkable. For as long as I can remember, the British farming ministry has seemed in the pocket of the National Farmers' Union, but not any more; in the wake of Brexit, the British government has charted a course to decouple agricultural subsidies from boosting production or owning land. Instead, farmers should be rewarded for providing 'public goods' – environmental and animal welfare benefits.

Gove criticised the EU subsidy regime, the Common Agricultural Policy, which has rewarded rich landowners for owning land rather than for good environmental practice. He went on to say that it 'encourages patterns of land use which are wasteful of natural resources and often intrinsically poor value, rather than encouraging imaginative and environmentally enriching alternatives'.[21]

It is ironic that Brexit happened just as the European Commission was setting its sights on reforming agriculture, including reducing pesticide and antibiotic use by half as well as increasing EU land under organic farming to a quarter by 2030. In the aftermath of the Covid-19 crisis, the EU Council declared industrial agriculture a factor in increasing the 'risk of future pandemics' and called for it to be 'tackled' as a global priority.[22] And the UN's former Special Rapporteur on the right to food, Hilal Elver, told the Human Rights Council that 'the current industrial agricultural model mistreats animals, emits greenhouse gases, relies on toxic pesticides, pollutes ecosystems, displaces and abuses agricultural workers and fisher folk, and disrupts traditional farming communities'.[23]

Attitudes to industrial agriculture have been shifting for a while. I remember attending the launch of a book called *More Human: Designing a World Where People Come First*, which called for a complete ban on factory farming. The author was Steve Hilton, former adviser of the then prime minister, David Cameron. The event was awash with government heavy hitters. Chatting to Cameron, he seemed keen to assure me he was doing something about the issue.

I've always said that it takes thirty years for big change to happen, and the changes now being talked about in farming have been a long time coming.

Industrial farming is largely based on investing capital, in the form of using chemicals, cages and machinery to maximise harvests of crops, milk or meat. It has meant specialisation and a focus on one particular type of product. Livestock and crop farming have become separated from each other, with animals largely moved from the land into industrial barns. Chemicals have replaced natural fertility-building and the management of weeds and insects, whilst machines have replaced people.

Thirty years ago, the true cost of intensive farming began to become obvious, in terms of its impact on animal welfare, the environment and wildlife. Industrial agriculture had become the predominant way of producing food, and policymaking seemed determined to defend it at any cost. In response, campaigners had to take a sledgehammer to a bolted door; reform was slow and those who called for it were seen as radical. The main way to get anywhere back then was to chip away at the most indefensible parts of the industrial machine, whether narrow crates for veal calves, battery cages for hens, hormone growth promoters or genetically modified food.

Now, things are starting to change. Cabinet ministers and senior politicians like Michael Gove and Zac Goldsmith have been advocating for a new direction in food and farming, one based on restoring nature's capital. Leading lights in mainstream farming have started to take matters into their own hands and change things. That the need for change is widely recognised is a cause for celebration. But with so many animals suffering confinement while soils and wildlife are ebbing away, the chances of genuine agricultural sustainability are shortening. Time is running out. The question is, can we act fast and far enough to stave off farmageddon?

5

CHAIN REACTION

BRAZIL: WHERE ONE THING LEADS TO ANOTHER

Dust clouds form beside the impenetrable Amazon rainforest and the noise is deafening. A bulldozer clanks through a clearing in the forest, its engine roaring. It pulls a heavy-duty metal chain originally designed for mooring ships. The chain runs like a fishing line from a reel before pulling tight, causing the vehicle to grind to a halt. There is a screech of metal as the constrained bulldozer turns ninety degrees to face the forest. At the other end of the chain is another bulldozer, and the two move in sync through the forest. Nothing escapes. Everything is brought down. No matter how old or stubborn, bushes and trees and everything in between are swept to the ground as if they were twigs. Onlookers laugh as another patch of the rainforest is destroyed.

This is *correntão* – 'the chain' – a controversial means of deforestation. It was long considered illegal but was recently authorised in the Brazilian state of Mato Grosso. Whole stands of trees can be brought down in seconds, before the remnants of ancient forest are cleared – first by fire, and then by cattle.[1]

FROM BIG BEEF TO BIG SOYA

There's nothing new about the expansion of cattle ranching in the Amazon rainforest, but I was shocked to see the firm hand of factory farming driving further destruction to grow soya. Existing cattle pastures have been put to the plough to grow crops for chickens, pigs, fish and dairy, much of which is exported to China, Europe and the UK.[2] Flush with cash having sold their fields, cattle ranchers are moving deeper into the rainforest and buying more land. And while the world focuses on the growth of cattle ranching in the rainforest – a dramatic symbol of deforestation – the fact that soya is the driving force behind it goes largely unnoticed.

It's a recent phenomenon, with rainforest clearance having traditionally stemmed from logging for high-value timber, and the newly opened areas then being cleared for farming. The pace of deforestation had long been driven by logging and a global clamour for commodity beef and leather. Lately, global demand for animal feed has caused a soya boom in Brazil.[3] Worldwide, 99 per cent of soya meal – the high-protein solid separated from the oil of the bean – is used for animal feed, with the majority being used to feed pigs and poultry.[4]

Massive soya monocultures first ravaged the grassland savannah of the Brazilian Cerrado. New soya varieties meant that the humid conditions of the Amazon were no longer a barrier to the crop's production. Industrialisation of the landscape stepped up with the introduction of GM strains of the crop that had been modified to withstand herbicides designed to kill all other plants.[5] Then came investment in infrastructure, with the construction of roads and ports to help get grain out of the country. In 2003, the agribusiness company Cargill opened a $20 million facility at the port of Santarém, where the Tapajós and Amazon rivers meet. Driven by the global demand for soya to feed the world's factory-farmed chickens, pigs and cattle, 2004 saw the second highest-ever amount of deforestation in the Amazon basin.[6]

In the same year, the Brazilian government announced plans to pave the BR-163 highway, a road over 3,000 kilometres long that would allow year-round access to the heart of the western Amazon.[7] In the 1970s, the road that became known as the 'soya highway' had been little more than a dirt road cut through the jungle. In the wet season, parts of it were muddy and impassable, and lorries would get stuck for a week or more at a time.[8]

As the demand for soya increased, so did land prices, and factory farming sparked a 'soya rush'.[9] Today, the crop occupies more than 33 million hectares in Brazil – an area the size of Malaysia – up from 13 million hectares in 2000.[10] The 'soya highway' stretches all the way from soya fields to the riverside export terminal. Thousands of grain trucks pass along it to the ports during harvest time. Soya is then transferred from trucks onto barges and, following a trip downriver, it is poured into ships' holds and dispatched around the world.[11] More than a fifth of Brazil's soya exports go to Denmark, France, Germany, Italy, the Netherlands, Norway and the UK.

With much of BR-163 now paved, and with the Brazilian government planning a railway that would run parallel to it, I was keen to learn more about the human and environmental costs of soya-fed chicken and pork on the Amazon rainforest.

DREAMS

Not everyone who lives on BR-163 in the heart of the Amazon biome is connected to agribusiness; some people, like Osvalinda Maria Alves Pereira and her husband Daniel, have other dreams. 'We wanted our own land, so we could survive on what is ours. To know that we produced the food with our own hands, our own sweat, is the most important thing', Daniel said.

Osvalinda and Daniel had long dreamt of having their own smallholding. They had previously had a land reform plot in northern Mato Grosso, a state that has become the heartland of Brazilian industrial agriculture. Osvalinda explained that they had stayed on their modest plot for three years, growing a variety of crops alongside some dairy cattle and chickens. However, two decades ago, the area became surrounded by big agricultural producers.

'We didn't want to sell – we wanted to stay and live off the land', she said. 'But they would pass the *correntão* – the deforestation system composed of two tractors linked by a huge chain that sweeps the forest and brings it to the ground – and then set the fallen logs on fire.'

Osvalinda was already suffering health problems that affected her lungs, and the fumes from the burning timber only made the problem worse. Soon, their crops were ravaged by fire, and they could stand it no longer. They did what so many small farmers along BR-163 have done: they caved in and moved on. Although it was twenty years ago, Osvalinda and Daniel remember it like it was yesterday. They packed all their belongings onto a motorcycle and rode more than a thousand kilometres north to another settlement project – a newly opened area of land that was part of a social programme in Trairão, rural Pará. They thought they'd left their troubles behind, but once again found themselves as smallholders in a sea of big producers sharking for land, and it was then that the threats and intimidation began. One morning in 2018, Osvalinda went to tend to her chickens and found two graves meticulously dug with crosses, one for her and one for her husband.

Osvalinda and Daniel fled to the city. Their plight mirrors that of other small farmers in the wake of the soya rush; big producers moved in, while they were forced out.

Sixty-year-old João Batista Ferreira is another small-time farmer who is contemplating moving away. He owns a sixteen-hectare plot in rural Belterra, the epicentre of Brazilian soya expansion in the Santarém Highlands. During his childhood, the area was thick forest. His land stands beside the Tapajós National Forest, a 550,000-hectare conservation area that is home to threatened species including white-bellied spider monkeys, giant anteaters and jaguars. Back then, wildlife could have wandered far and wide in endless habitat, but not any more.

Ferreira's plot is an island of shade and birdsong in the middle of sweeping prairies of soya. His father taught him to be a beekeeper when he was nine, the start of a passion that would last for forty-five years. But when heavy pesticide usage and crop monocultures on surrounding land took its toll, his beloved bees went into decline.

Ferreira is still known as 'João of Honey', though few of his thousand beehives remain. Now, surrounded by industrial crops, he constantly thinks about selling up and moving deeper into the jungle to escape the soya siege. He talked about how agribusiness caused the destruction of native forest, and how it created so few jobs, all the work being done by machines. In an act of lonesome protest, he painted altered versions of the Brazilian flag with question marks rather than the national motto 'Order and Progress' – he wasn't sure Brazil had them any longer. 'One day progress comes', he said, 'and decay arrives with it.'[12]

For now, Ferreira is staying put but moving on seems inevitable. He still welcomes visitors and proudly shows off what's left of his simple wooden hives, yet he hates talking about what is happening. 'I have lost faith [in speaking out], that it can change anything. The more I speak of it, it seems the closer they [the soya producers] get.'

For some, like Osvalinda, Daniel and Ferreira, moving on becomes the only option. For most indigenous people living on ancestral land deep in the forest, moving on is unthinkable. Yet towards the end of BR-163, on the outskirts of Santarém, things are so bad that the indigenous Mundurukú people are starting to consider it.

FADING FOREST

I found members of the Munduruku people keen to show just what was going on and how their rainforest home was shrinking. Three tribesmen were anxiously investigating rumours that someone had been cutting down the forest. They wanted me to see what was going on too. Along the way, they offered Brazil nuts from the forest floor, encased within a hard shell. With each fallen pod weighing around a kilogram, we counted ourselves lucky that we were tasting the forest's produce rather than being hit on the head with it.

After an hour of walking, light started to shine through the dense vegetation before the forest opened out into a blanket of soya bushes. For at least a kilometre, there were no trees; beyond the crops, the forest rose again. Only, something didn't look right. The tall forest canopy was intact, but the smaller trees had been felled. This was selective logging, a practice employed to avoid satellite detection by preserving the canopy. The rumours were true. Whether or not the satellites had noticed, another patch of the Munduruku's forest had disappeared.

'It seems they are using tractors to remove the smaller trees, while keeping the bigger ones. Come summer, they'll set it on fire,' Paulo Munduruku said, his voice full of resignation. He'd seen it all before. 'Deforestation advances every year: someone deforests here, someone else deforests there. And the forest ends up fading.'

The remaining trees would likely fuel the next 'fire season', a period when the Amazon burns with fires started by people keen to clear the land for farming. The previous year had seen more than 2,500 major blazes across the Amazon, some within conservation areas and indigenous reserves.[13]

Despite losing parts of their forest to crop production, Paulo and his companions were a long way from giving up. 'This is what keeps us standing. To see our forest blooming, alive, for our survival, the survival of the animals, the birds. This is why we claim justice.'

Paulo, Manoel and Lino wore T-shirts and rubber boots, while Paulo also sported a necklace made of seeds and an animal's tooth, a sign of their tribe's history as fierce forest dwellers. The Munduruku had a reputation as one of Brazil's most powerful tribes. They were once known as the 'head cutters', a reference that no doubt struck terror into their enemies. Today, they are peaceful, seeking instead to 'cut off the head' of enemies through dialogue.[14] Yet, nothing in their history could have prepared them for going head-to-head in dialogue with agribusiness.

The disappearing forest of his ancestral homeland is what keeps Manoel Munduruku awake at night. 'My biggest concern is the land-grabbing, the soya crops, the deforestation', he said. He has been leading his people's claim for recognition of their ancestral rights for more than decade, but progress has been painfully slow. The attorney general's office in Santarém launched a lawsuit in an attempt to push back against agribusiness. 'The lack of action [by the government] has transformed the territory into the epicentre of a series of rights violations associated with the soya monoculture', it reads, before going on to list the threats to the territory: 'Deforestation, destruction of archaeological sites, destruction of bodies of water, air, fauna and flora contamination by pesticides, land-grabbing attempts, threats and intimidation, among other problems.'[15] However, Covid-19 brought proceedings to a grinding halt.

'All of this is killing us silently', Manoel said tearfully. 'No money in the world is worth what this land is worth. I learned from my parents that it is shameful to shed false tears. But when you are fighting for what is right, there is nothing more noble than to cry.'

LAND-USE CASCADE

Where the 'soya highway' enters the state of Pará, the change is particularly marked. The municipality of Novo Progresso is gripped by logging, mining and agribusiness. Both sides of the road are lined with shops and businesses selling related products, and I was reminded of Mato Grosso, a previous soya rush frontier. Stainless-steel grain silos were dotted around, giving the place the feeling of a rapidly expanding agro-industrial outpost.

Cattle ranching, which has long been a fixture in this area, is being replaced with soya in what has become known as a 'land-use cascade'. This cascade effect, triggered by demand from the soya industry, has caused land prices to rocket by as much as ten-fold. Cattle ranchers have been able to sell their fields for enormous profits and buy land further north to expand their herds.[16]

In Novo Progresso, the authorities are open about this soya-sparked chain reaction. With the help of a local journalist, I interviewed the town's agriculture secretary Cleiton Júnior de Oliveira, a cattle rancher himself. He said, 'I can confirm that the cattle are being pushed to the extremes of the forest. Plains areas with great agriculture potential are turning into crops, and occupying the space previously held by cattle. It's a fact.'

In this way, expanding soya production is driving cattle deeper into the Amazon, causing yet more deforestation. American scientists have found that a 10 per cent reduction of soy expansion into existing pasture areas between 2003 and 2008 would have reduced deforestation in the Amazon by as much as 40 per cent.[17] This reflects rising land values and the fact that cattle farmers who have sold their land can buy so much more land elsewhere.

In 2018, Transparency for Sustainable Economies (Trase) described how soya drives the indirect conversion of forest and savannah by increasing land prices, leading to speculative clearing, with cattle ranching being moved further into the forest.[18] Trase noted that crop expansion is playing a 'key, indirect role in driving deforestation in Latin America via the displacement of cattle pasture'.[19] In that year, the expansion of pasture was responsible for four-fifths of the deforestation in the Brazilian Amazon, over 95 per cent in the Paraguayan Chaco and more than half in the Cerrado.

The International Bank for Reconstruction and Development, part of the World Bank Group, has reported the devastating effects of the land-use cascade thus: 'The high profitability of soy production on suitable Amazon soils has pushed land prices upward, providing new sources of capital to cattle ranchers who sell or rent their land to soy producers.'

The report focused on the Brazilian Soya Moratorium, an agreement designed to protect huge swathes of Amazon rainforest. Soya expansion in the Amazon, it said, had 'not been suppressed' by the moratorium, 'because of the abundance of suitable cattle pastures for conversion to soy'. But deforestation for cattle ranching was clearing the way for the subsequent expansion of soya, circumventing the moratorium on deforestation for soya.

The International Bank for Reconstruction and Development had no choice but to admit that the battle was being won, but the war lost. The moratorium had ensured a deforestation-free soya sector in the Amazon, but the trend towards greater forest clearance had continued.[20] Essentially, cattle had cleared the forest first to allow subsequent 'deforestation-free' crop production.

I spoke to Mauricio Torres, a professor at the Federal University of Western Pará, who blames the land-use cascade on the global demand for soya. 'It is an indirect process. The soya advances over the pastures, so the cattle rancher fills his pockets and heads to deforest the Amazon.' With the paving of BR-163 and improved port facilities now in place, 'it is much cheaper for the soya producer to export. Hence the impacts can be traced back to the soya industry', Torres said.

The prevailing political and economic climate in Brazil seems likely to cause more deforestation. Soya production is more lucrative than cattle ranching, providing double the income per hectare. Opening up the forest for cattle pasture has become a way in which ranchers can gain control of large swathes of land while waiting for new infrastructure and higher land prices.[21]

The Brazilian president Jair Bolsonaro, having come to power with the support of farmers, truckers and miners, was keen to keep his promise of developing the world's largest tropical rainforest. With BR-163 now fully paved, the Brazilian authorities were looking to double-down on their newly created global artery for soya exports by building a parallel railway, a $3 billion 'grain train' to run alongside.[22]

FIRE DAY

In August 2019, surging fires in the Amazon captured headlines around the world, prompting criticism of Bolsonaro's government for not doing enough to protect the rainforest. The following year, satellites recorded 32,017 fires in the world's largest rainforest, nearly two-thirds more than the previous year.[23] Most fires in the Amazon are not natural occurrences, but man-made. Blazes are thought to have been set deliberately by ranchers and land-grabbers keen to expand agricultural operations and cattle grazing. The fires sparked a war of words with the then US presidential candidate Joe Biden, who threatened to impose 'economic consequences' on Brazil if it did not 'stop tearing down the forest'. Bolsonaro described Biden's comment as a 'cowardly threat' to his country's sovereignty and a 'clear sign of contempt'.[24]

The city of Novo Progresso found itself the subject of international headlines because of what became known as 'fire day'. Local ranchers, land-grabbers and businessmen along BR-163 are believed to have marshalled hired hands to start the fires, while others tried to suggest that NGOs had caused them to gain attention. The number of fires increased by 300 per cent in two days.[25] Reuters reported that organisers were suspected of coordinating the fires via WhatsApp as a show of public defiance against environmental regulations, describing it as a 'coordinated invasion to force the area into farmland'.[26]

Novo Progresso sits beside the Jamanxim National Forest, a protected reserve of more than 1.3 million hectares. Commercial agriculture is not permitted, a fact that many local people have ignored.[27] Farms in the area were once going to be repossessed by the government, but that never happened, and wildlife has struggled in the wake of farmland expansion. A *Guardian* reporter who visited the region was struck by 'how farming had eaten into the forest on both sides of the BR-163. Cows were everywhere. Wildlife survived as best it could. One morning a white monkey scuttled across a dirt road, followed by a gaggle of forest pigs. Black, blue and orange macaws squawked atop a charred tree trunk, their only perch in a field of cattle. An opportunistic anteater darted across the highway in a gap between the trucks.'[28]

Here, as elsewhere in the Amazon, deforestation has followed a set pattern. First, loggers remove the most valuable trees, leaving cover that hides the damage from satellites. The remaining trees are then felled, left to dry and burned, before grass is sown and cattle are put on the land. And then comes soya.

Novo Progresso was first developed thirty years ago, when Brazil's military dictatorship lured families to the area with the promise of land and opportunity. The armed forces, where President Bolsonaro began his career, viewed the largely uninhabited Amazon as a vast, resource-rich asset. But when the dictatorship fell, the democratic government that replaced it followed a very different policy: conservation.

The early pioneers in the Amazon, many of them cattlemen, felt unsupported and betrayed until Bolsonaro took over. Novo Progresso's influential president of the rural producers' union, Agamenon Menezes, was interviewed by police in connection with the 'fire day' investigation, but denied any involvement.[29] In his view, the expansion of crop production is inevitable. 'Soy will have to come in … crops are paying out better than cattle, the flat areas are being turned into crops. Cattle remain in the hilly areas,' he said.

Like other farmers of his generation, Menezes has little time for environmental concerns. 'If what the monitoring agencies and NGOs said about deforestation in the Amazon was true, there would be no trees left around here. The Amazon would be gone. We don't deforest – what we do is replace native vegetation with another kind of vegetation, the crops.'

Yet the continued rollback of policies designed to protect the forest have enabled higher deforestation rates.[30] By late 2019, an area of rainforest the size of Devon and Cornwall had been lost in the Brazilian Amazon in just twelve months, the highest rate of loss in a decade.[31] In just three months of the following year, three hundred square miles of Amazon rainforest were destroyed.[32]

With deforestation comes conflict over land.[33] Mariana Rodrigues is president of the settlers' association of the sustainable development project, PDS Brasília in Castelo dos Sonhos. She lives in the shadow of what happened to Bartolomeu Moraes da Silva, brutally murdered in 2002 for trying to protect his people's land.[34]

For much of her life, Rodrigues worked on the land. However, when an accident ended her career, she got involved in the politics of settlements, newly opened areas of land. But this put her at odds with cattle ranchers who were setting up in areas meant for smallholders. She calls the ranchers 'concentrators', collectors of land.

Legally, the plots of land in a settlement are not supposed to be commercialised, but Rodrigues says cattle ranchers try to take them over by wearing the settlers down, hampering access to their plots or allowing cattle to trample their crops. A settler's house was recently run over with a tractor.

Exhausted and frightened, the settlers often cave in and sell their land for whatever they can get – they just want to leave. While individual plots tend to be around twenty hectares per family,

Rodrigues says that one 'concentrator' alone had amassed over five hundred hectares.

Rodrigues has received threats herself and is at the end of her tether. 'Our hands and feet are tied. We reach out to the police, but they do nothing. What ends up killing people in these conflicts is the judicial system, being so slow. My family is really scared. The concentrators send messages through other people, telling me to watch out.'

While no soya is currently grown on the land-grabbed plots inside the settlement, Rodrigues thinks it won't be long. 'These lands are really coveted. They are all flat plains, good for crops.' Her story reminded me of the graves that were dug for Osvalinda and her husband. After the incident, the police offered to place the couple in witness protection, but they declined because it would mean they could never come back. Poor health caused them to think again, and they reluctantly left their beloved plot in the hands of a housesitter. Despite being away for over a year, the intimidation continued: rumour had it that they'd be killed if they ever went back. However, they found life in the city suffocating and eventually returned home to find their crops overgrown and their chickens had been stolen. 'But it was like being born all over again', Osvalinda said.

Against all odds, the couple renewed their simple dream: to have a patch of land and live off it. Some people threaten to kill for another's land, but they are willing to die for theirs.

The lives of ordinary people, animals and the finely balanced ecosystem are all under threat in the Amazon. Clearing the odd patch of a vast forest may seem innocuous. Yet, when one thing leads to another – logging, then cattle, then soya – things can quickly get out of hand. *Correntão* sets off a chain reaction that leads to soya. With a sixth of the Amazon rainforest already lost, the fear is that a tipping point could soon be reached where the entire rainforest could dry out and degrade into savannah.[35] And with it would go the means for preserving nature and slowing climate change. While cattle ranching has long been linked to the deforestation of the Amazon, it has become clear that the driving force behind it is closer to home: the cheap soya-fed meat on our supermarket shelves.

6

A LAND WITHOUT ANIMALS

IN SEARCH OF SKYLARKS

Abruzzo, Italy, and the views were stunning, as finches, titmice and other songbirds twittered delicately in the cool of the morning. From my balcony, I could see the peaks of the Apennines rising imposingly, their torsos covered in pine trees. Sunshine rested gently on verdant slopes as I waited to be picked up for a day in the hills.

A group of officials from the Gran Sasso National Park were keen to show me how the area benefits from the restoration of cows and sheep as free-roaming grazing animals. Up here in the mountains, I was told, nature and food production worked hand in hand.

We were soon bumping along in the park ranger's Land Rover, as we climbed above the treeline and past a small herd of tan-and-black cattle that were grazing on lush pasture punctured by gleaming white rock. When we stopped, I could hear the delicate rise and fall of the skylark's song, as the birds hovered high in the sky and fluttered down like an autumn leaf.

The skylark's trill has long been a familiar serenade in spring and summer, and more poems have been written about its song than that of any other bird. It has moved writers including George Meredith ('The Lark Ascending'), Ted Hughes ('Skylarks') and Percy Bysshe Shelley, whose 1820 poem 'To a Sky-Lark' was inspired by hearing their song in the Italian countryside. Shelley's wife Mary wrote of him noting the 'carolling of the skylark' as they wandered among 'the lanes whose myrtle hedges were the bowers of the fire-flies'.[1]

During my time in the Apennines, I also felt inspired by the skylark's song; that and the coolness of the fresh mountain air, a blessed relief from the sweltering heat I had experienced while travelling through the agricultural heartland of the Po Valley. The Apennines and the Po Valley could not have been more different. One had wildlife and farming working hand in hand, while the other was a land without any roaming animals. The Po Valley is an expansive area. Over four hundred miles long, the River Po has 141 tributaries fed by the Alps and the Apennines. More than 16 million people – nearly a third of Italy's population – live in this fertile lowland basin that stretches from the French border to the Adriatic Sea.

The Po region, which is home to some 3 million cattle, between 5 and 6 million pigs and an estimated 50 million poultry,[2] is renowned for its world-famous cheeses and its Parma ham. As a lover of Italian food, I was excited at the prospect of spending a few days exploring the beauty of the Italian countryside, perhaps also sampling a little wine and some of the local fare. However, I soon discovered that all in this gastronomic land is not quite as it seems.

***

Before reaching the Apennines, I'd travelled through the Po Valley with Annamaria Pisapia, who runs the Italian office of Compassion in World Farming in Bologna. An effervescent woman with sparkling eyes and a big smile, Annamaria was keen to show me around. 'I'm taking you on a tour of one of Italy's most famous regions. You'll see the countryside with fields of grass, but not as you know it,' she told me. It was late June. We passed through amazing cities and quaint towns. The scenery was stunning. Much of the architecture was simple with painted walls, all bleached and faded by relentless sun. I marvelled at a spectacular walled castle with towers and fluttering flags, cast against a hillside covered in grapevines. The houses in the valley below had roofs made of characteristic terracotta tiles.

Of all the countries in continental Europe, Italy is the one I feel most at home in. I love the food, the friendliness, the countryside, the pace of life and the tradition. Yet one thing rankled with me during my trip: I had travelled for days without seeing a single farm animal in the fields. We had been driving through a region that was famed for its cheeses, but where were all the cows?

When I checked the production details for Parmesan and Grana Padano, they both mentioned the cows being raised on fresh grass and hay.

There was plenty of grass, some growing tall in fields with stalks that rippled in a welcome moment of breeze; in other fields, grass had been cut and lay in tightly rolled bales like golden curls of butter. And as you would expect in a fertile valley, there were plenty of crop fields.

But cows wandering the fields, gently tugging at grassy stalks or chewing the cud under the shade of trees, were conspicuous by their absence; the only cows I could find were crammed into darkened sheds.

A hundred cows peered out of a single square shed. They were trapped inside but had a perfect view of the lush pasture they were missing out on. I later watched, incredulous, as heavy goods vehicles delivered grass to a processing plant. A yellow tractor scooped grass into a giant machine, while yet more trucks took compacted blocks of grass away. Nature's way had been abandoned in what should have been a cow's paradise. Instead, confined animals were being fed mechanically pressed bales.

As we approached the city of Mantua, a road sign indicated that we were in the home of one of the region's speciality products: Grana Padano cheese, which is produced all over the Po Valley and is said to be the most important cheese in Italy. First created by the Cistercian monks of Chiaravalle Abbey as far back as the twelfth century,[3] it is milder and less crumbly than its long-aged sibling, Parmesan.[4]

Grana Padano is made to strict specifications.[5] Only cows that are fed a diet of 'fresh forage from meadows', ideally including a mix of meadow grasses, alfalfa and clover, can provide the milk to make it. Surely, I thought, these cows would live outside, where they could enjoy the fresh air and sunshine?

Dr Stefano Berni, general director of the Grana Padano consortium, burst my bubble when he told me that only 30 per cent of the cows graze naturally, while the rest are confined indoors. When I asked him whether the specifications cover animal welfare standards, his response was blunt: 'It is not provided for.' I was shocked.

When I investigated further, I found that 'zero-grazing' – keeping cows indoors permanently – is common all over Italy. As Kees de Roest wrote in *The Production of Parmigiano-Reggiano Cheese: The Force of an Artisanal System in an Industrialised World*, 'almost all Italian dairy farms apply the zero-grazing system'.[6]

I couldn't help but notice that Grana Padano's rules allow cows to be fed ingredients like cereals and soya, things that could otherwise feed people directly.[7] Cows are ruminants, their four stomachs allowing them to live on grass and other herbage. Rearing them on a natural diet allows them to turn something people can't eat – grass – into milk or meat, which they can. However, as I've seen all around the world, intensive farming turns this simple logic on its head. Cows are shut away and fed grain-based concentrates, thus creating competition for food between cows and people.

So I now knew that most cows whose milk is used to make Grana Padano are kept indoors permanently, but what about those milked for Parmesan? To connoisseurs of good cheese, authentic Parmesan – or Parmigiano Reggiano to give its full name – is one of the best. Surely, this 'king of cheeses' would be made from the milk of cows that are kept on grass?

Like Grana Padano and many other regional foods, strict rules govern the manufacture of Parmesan.[8] Only cheesemakers from the provinces of Parma, Reggio Emilia, parts of Bologna, Modena and Mantua can qualify, given its protected status. The most famous of all the hard 'Grana' cheeses, Parmesan is made from unpasteurised cow's milk, from animals that are 'mainly fed by the local forage, grass and hay'.[9] Surely, then, the cows should be out grazing, at least in the summer? Dr Marco Nocetti from Consorzio del Formaggio Parmigiano Reggiano, the consortium that governs the manufacture of Parmesan, was quick to dispel the idea. 'The percentage of cows grazing freely in fields during the summer months is very low', he told me.

It turns out that nearly all cows that are milked for Parmesan are permanently housed in stalls, with forage taken to them. 'There are few exceptions – around 1 per cent of the farms – and they are located mostly in the mountains or on the hills', Nocetti said. In Italy, agriculture is almost exclusively intensive, which Nocetti puts down to a lack of land. 'The territory available for agriculture is relatively little compared to the population.'

Yet Parmesan producers do place great importance on cow nutrition. Nocetti told me about a rule that 'every day the cows have to be fed with a bigger portion of forage than feed. This is fundamental because cows are ruminants, and this kind of feeding respects the physiology of the cow and her rumen.'

Then why not let them out to graze instead of mowing the fields and bringing the grass to them? The simple order of nature seems to have been lost in this part of Italy. Most of the nation's dairy cattle[10] spend their life indoors and have grass brought to them, instead of grazing the lush fields of the Po Valley. As elsewhere, crop fields are often dedicated to growing feed for cows rather than food for people. About half of Italy's cereal crop is used to feed intensively reared animals. In addition, the country is a big importer of soya, importing over 3 million tonnes a year,[11] much of which is used to feed industrially reared animals.

My conversations with the consortia for Parmesan and Grana Padano troubled me to the point that Compassion in World Farming's Italian branch investigated their treatment of zero-grazed cows. We found out that, far from the image of hard-working artisanal creatures shown in advertising, the cows were treated like milk machines. Investigators found them with their bones protruding, all too often with bleeding open wounds and lying on concrete in their own muck.[12] It led to the Parmesan consortium carrying out a welfare assessment on their suppliers' farms and introducing voluntary measures that included financial support for improvement. The measures fell short of what we saw as acceptable animal welfare – they didn't even forbid tethering cows in the barns, let alone ensuring they could enjoy access to pasture. And Grana Padano's response was even less convincing.

It is not only dairy cattle that are affected; in Italy; most of the country's beef cattle are zero-grazed, often spending their lives in grassless paddocks until they are carted off for slaughter. I found it difficult to understand why such beautiful countryside was so devoid of farm animals.

A lifetime spent indoors can't be much fun for the cows, but how does it affect other countryside dwellers like skylarks? I called in to the Parma office of Italy's main bird protection organisation, Lega Italiana Protezione Uccelli (LIPU), and met Patrizia Rossi, farmland bird officer. The office walls were adorned with framed paintings of old zoological atlases. Shelves were stacked with painted nesting boxes and bird feeders.

I asked how farmland birds were faring in the Po Valley. 'The situation here is probably the worst in Italy', Rossi said. Once-common farmland bird species seemed to have hit rock bottom at the end of twentieth century, yet they have continued to decline. In the fifteen years to 2014, numbers had fallen by 18 per cent across the country. In areas of intensive agriculture, numbers have fallen by as much as 40 per cent in the same period. Skylarks had been particularly hard hit, declining by 45 per cent.

Grasslands have been ploughed to make way for intensively cultivated crops like cereals. Furthermore, the remaining grassland is often managed intensively, being laced with fertilisers and cut frequently so birds have little opportunity to rear their young.

I started to understand how the fates of skylarks and the region's cows were interrelated. 'We know that birds in the Apennines and in the south are doing better,' Rossi said, 'probably because the agriculture is less intensive.'

I had travelled widely in the Po Valley without seeing a single skylark or cow in the fields, and now I knew why. I discovered that, in this land without animals, much life had been lost from a once-thriving countryside. And the route to getting it back lies in restoring animals to the land.

INVISIBLE PIGS

The city of Parma was surrounded by fields of ripening wheat, barley and maize. Remnants of farm buildings crumbled in the sun, proof of the region's long-standing agricultural heritage.

I passed a road sign that indicated I was entering a 'UNESCO City of Gastronomy'. It was adorned with an illustration of a joint of pork, representing the Parma ham for which the region is famous. However, though I kept an eye out for pigs in the fields, I saw none.

Eventually I came across a traditional yellow brick building in Mantua province that turned out to be a factory that produced ham. The place was shut, so to get an idea of the rearing conditions I had to rely on the findings of a previous undercover investigation by Compassion in World Farming. After visiting eleven farms in the region in 2013, investigators returned with footage showing pigs living in barren conditions with no straw bedding material, in breach of European animal welfare rules. Many pigs were filmed in squalid, cramped conditions with no space to escape their own waste. The tail of every animal had been docked – most likely without pain relief, despite EU rules that forbid the practice.[13]

One of the investigators gave an eyewitness account of conditions: 'I'd never seen pigs so deprived of stimulation, that the only thing to occupy their inquisitive nature is to play with their own waste.' They described seeing pigs cowering in dark corners, bullied by their pen mates and trying to escape the filth of their surroundings. Similar conditions have been documented by others, not least by the journalist Giulia Innocenzi, whose findings on living conditions in farms certified for Parma ham were aired on primetime Italian television.[14]

That evening in nearby Mantua, the provincial capital of Lombardy, restaurants served diners on cobbled streets where waiters grated cheese over pasta dishes and loitered before serving the next gastronomic delight. Amid the chatting and clinking of glasses, I looked up to see hams dangling in neat rows from restaurant beams, a symbol of elegance, culture and of what seemed to me to be a questionable tradition. How much of Parma ham's place in gastronomic folklore is owed to the majority of the region's farm animals being kept out of sight, out of mind?

SOMETHING IN THE WATER?

It was like standing between two different worlds. On one side, the River Po was lined by grasses, willow trees and wildflowers; on the other, a low plain of monotonous crops was broken only by the jarring promontory of a warehouse rising up from the valley floor. This was a hen farm, I was told, with hundreds of thousands of birds, perhaps kept in cages.

It wasn't quite the introduction I was expecting to the Po Delta, Italy's premier wetland, where the river meets the Adriatic Sea. Here, I discovered that intensive farming affects more than the region's dairy cows and skylarks: the quality of drinking water is under threat, too.

Bestowed with meltwater from the Alps, the delta forms a vast mosaic of wetland areas which, along with the nearby 'Renaissance city' of Ferrara, is recognised as a UNESCO World Heritage Site.[15] The Po Delta Regional Park is a sprawling celebration of wetlands flanked by Mediterranean scrub, woodlands, reed beds and dunes. Danilo Trombin, a local guide, grew up around here and knows the landscape well. 'When I was a little child,' he told me, 'my grandfather took me with him along the river, fishing, looking at the trees, plants and animals. I grew up along the river and now the river is inside me.'

I asked him about the wildlife. 'All the herons breed here', he said, referring to purple, grey and squacco herons, as well as little, great and cattle egrets. Spoonbills are here, too. 'The river is increasingly popular with kingfishers.' What about skylarks? 'We get some in this area, but they are declining due to the intensive farming environment.'

During Trombin's lifetime, the countryside along the Po has changed dramatically. Much of it is now used for industrial agriculture: animals have disappeared from the land while crops are grown in chemical-soaked monocultures.

'These monocultures are like a green desert,' he said despondently. 'The only type of life here is the one being grown: the crop. The riverbank here is like the border between two worlds.' However, as I was soon to discover, one 'world' was starting to have a big effect on the other.

Due to pollution from agriculture and other sources, the fragile ecosystem of the delta is facing another threat: algal blooms. Most years, during the hot summer months, there is a green tide of algal bloom that can be lethal to aquatic life. Boosted by fertiliser run-off from farmland, algae multiply in vast quantities and then die. As they decay, they suck the oxygen out of the water, creating a dead spot where little can survive. 'With less oxygen, many fish and clams die, and there's much damage done to the local economy, which is based on clam fishing', Trombin said. It also gives off a pungent smell of sulphur. 'When the wind changes direction, visitors get a faceful of the smell.'

Trombin's concerns are shared by Professor Pierluigi Viaroli from the University of Parma, who spent a decade studying the delta's 'green tide'. From March to mid-May, the lagoons at this renowned wildlife site would see 'very large green sheets appear, or long filaments, floating on the water and growing very quickly', Viaroli told me.

The problem was causing the growth of a 'dead zone' along the coast of the Emilia-Romagna region. Typically found in the ocean but occasionally in lakes and even rivers, dead zones are areas of water with insufficient oxygen to support marine life. The cause of such 'hypoxic' conditions is usually an increase in nutrients in the water, which leads to oxygen-depleting algal blooms. Nitrogen and phosphorous from agricultural run-off are the primary culprits, but sewage and industrial emissions also contribute.

The professor had no doubt where most of the Po watershed pollution was coming from: 'agriculture and livestock', though he was careful to point out other factors such as sewage outflows and wastewater processing. I mentioned visiting the Gulf of Mexico, where a dead zone the size of Wales wipes out almost all life. 'We are having a similar situation in some lagoons and coastal areas in the north Adriatic Sea,' Viaroli said.

Dead zones are the greatest environmental challenge to coastal areas, and the Adriatic Sea is more affected than anywhere in the Mediterranean. Severe oxygen deficiencies affect as much as 4,000 square kilometres.[16] Scientists have been studying how animals respond to the sudden drop in oxygen. As Michael Stachowitsch from the University of Vienna has explained, 'Some try to escape and make their way up to higher water strata, while others curtail their activity or display entirely unnatural behaviour.'[17] Mass mortalities normally result.

However, coastal dead zones are not the only environmental threats affecting the region; drinking water is being contaminated by chemical pesticides.

Pietro Paris works for the Italian Institute for Environmental Protection and Research, which has found that contamination from chemical pesticides is widespread in the Po Valley. A survey of 1,035 groundwater sites and 570 surface water locations found pesticide contamination at more than 70 per cent of surface water sites and more than 40 per cent of groundwater locations. Of these, pesticides were found to be above the legal 'safe' limit in 32.6 per cent of surface water samples tested and 8.7 per cent in groundwater.

The situation is 'quite serious', Paris said. I asked him about the cause. 'We have a high load of pesticides because there is intensive agriculture', he said, before going on to explain that even where contamination levels are considered low, the substances involved can be lethal. Even more worrying is the discovery that pesticides are present in deep underground aquifers, where they can be 'preserved' for long periods of time. It means that substances that are banned or withdrawn from the market can be found in the environment decades later, where they can mix with other chemicals in potentially dangerous cocktails.

Paris estimates that in the 1990s, around a thousand different chemical pesticides were licensed for agriculture in Europe, many of which are now banned. 'Substances no longer on the market can represent a problem because of the environmental persistence.' Atrazine is one example; banned in Italy since 1991,[18] it remains a 'major contaminant' in groundwater in the Po Valley.

The situation is not unique to Italy. 'It's not completely different from other countries that have intensive agriculture and farming. We know for example the French situation … is similar', Paris said. So, what can be done about it? Paris believes we should be looking at ways to minimise pesticides in the environment.

My journey through the Po Valley had demonstrated the damage that industrial agriculture does to the countryside, its farmed animals and wildlife, and I had discovered that drinking water is also affected.

I was keen to see if there was another way, and that was how I ended up bumping up mountain tracks in the Gran Sasso National Park. I was in a Land Rover with the park ranger and a vet as we followed the pylons of the cable car up the mountain road. Pale green pasture covered with purple and yellow wildflowers stretched ahead of us. Sometimes the grass gave way to pale seams of bare rock. As shadows from clouds floated across the ravine like film credits moving over a picturesque screen, I learned how wolves and viper snakes make this area their home, along with wild boar and mountain birds such as the sleek black alpine chough with its bright yellow bill.

We passed two wooden shacks that resembled something out of an old cowboy movie. These were mountain-top restaurants offering local specialities to the visitors who flock to this peaceful wilderness.

Our Land Rover juddered towards two white tents, with several large white dogs waiting outside them. This was a milking station for sheep, and the resulting produce would be used to make pecorino cheese. I got out of the car and watched as sheep released from the tents skipped past me. They were being ushered out by 58-year-old Giulio Petronio, who has farmed in this area for his whole life. He makes Canestrato cheese, a type of pecorino, from the milk produced by his mountain herd of nomadic sheep.

The park is home to some 70,000 sheep, which belong to about two hundred farmers. Around 8,000 cattle, mainly of a more natural dual-purpose breed, calve up here, too. Small fields of barley were surrounded by wildflowers, red poppies and purple vetch. The park's vet, Umberto Dinicola, told me about the benefit of grazing: it produces better quality produce while also keeping the prairie alive and boosting the biodiversity of the area.

I watched as a wild boar, with black Mohican hair along his back, trotted through a carpet of wildflowers and disappeared into the scrub. All around me, grey cattle with long horns and bells were suckling their calves. And in the sky above, tiny black specks made a bubbling trill: skylarks.

In the lowland Po Valley, I had seen a land without animals. But here, I could see – and hear – something profoundly different: a taste of what the countryside could be like.

7

CLIMATE CRISIS

WHY PIGS AND POLAR BEARS DON'T MIX

Winter in Novaya Zemlya, a Russian archipelago in the Arctic Ocean, and a snow-covered rubbish dump had been ransacked. Wooden crates had been tossed aside and cardboard, cans and plastic packaging had been scattered in search of something valuable. Anyone who came near was chased away, and children were frightened. 'People are scared, afraid to leave the house ... afraid to let their children go to school', a local administrator was reported as saying. Officials on this remote island, which is mainly populated by military personnel, declared a state of emergency. In decades of living here, locals had never seen anything like it: the island had been invaded by polar bears.[1]

According to reports, as many as fifty-two polar bears were rummaging through garbage in search of food. Adults and youngsters alike jostled for what edible pickings they could find, getting into dumpsters and even entering buildings. Extra fences were erected around schools, an attempt to keep the hungry visitors out. Special patrols attempted to shoo them away.

When the bears found something worth eating, a mass of thick tan-white coats would converge. Those animals who were unable to get to the action glanced around anxiously, their darkened snouts twitching. In the middle of the pack, the odd angry scuffle broke out, a muffled roar accompanied by a coming together of heads as they competed for scraps.

All of this on an island territory known more as a nuclear testbed. It was here in the 1960s that Tsar Bomba, the world's most powerful nuclear explosion, was detonated. Now, this remote island is witnessing the real-life consequences of something no less dramatic and more far-reaching: climate change.[2] Arctic sea ice is shrinking by 13 per cent each decade,[3] making it harder for polar bears to find food naturally. They would usually feed on seals, walruses and whale carcasses,[4] all of which provide bigger meals than scraps from a garbage dump.

Commenting on the polar bear invasion, WWF Russia said, the 'number of human and predator encounters in the Arctic is increasing. The main reason is the decline of the sea ice area due to the changing climate. In the absence of ice cover, animals are forced to go ashore in search of food, and settlements with spontaneous waste deposits are the most attractive places.'[5]

Real-world examples of accelerating climate change are increasingly catching the eye of those in positions of power, but are they prepared to do what's necessary to save the planet?

TIME RUNNING OUT

'Code Red for humanity' is how the UN Secretary-General described the latest scientific warnings about climate change in 2021. 'The alarm bells are deafening, and the evidence is irrefutable', said António Guterres, with the effects impacting every region on Earth, putting billions of people at immediate risk, with many changes becoming irreversible.[6]

As time continued to tick away in what scientists have suggested could be the final decade to avert climate catastrophe, official warnings have become ever more stark. Global reporting by scientists from the Intergovernmental Panel on Climate Change have found that limiting global warming to 1.5 degrees Celsius would require 'rapid and far-reaching' transitions, including in energy, transport and land, with the latter – land – being shorthand for agriculture.

An increase of half a degree more could mean drought, floods, extreme heat and poverty for hundreds of millions of people.[7]

Though the link between those dump-invading polar bears and the rearing of pigs, chickens and cows for food may be far from obvious, it will be critical if we are to stave off the worst effects of climate change. Food is responsible for between 21 and 37 per cent of global greenhouse gas emissions.[8] In agriculture, the production of animal products is responsible for up to 78 per cent of all emissions.[9] As it is, the livestock sector produces more greenhouse gases than the direct emissions from all forms of transport.

A paper published in *Nature* in 2018 showed that on current trends, greenhouse gas emissions from food production and consumption would increase by 87 per cent by the middle of the century. The bottom line is that more people eating more meat from resource-intensive industrial agriculture makes for a hothouse planet. Those same factors would also increase demand for cropland and water by two-thirds and fertiliser use by more than half.[10]

In many discussions about climate change, cows are singled out as a big problem because of the methane they emit, while factory-farmed pigs and chickens are overlooked. Industrial pigs and poultry may not emit large quantities of methane in the same way as ruminant animals, but their rearing still produces serious emissions. Carbon dioxide is released from the intensively managed soils needed to grow their feed. In addition, intensively reared pigs and poultry are fed soya from farmland in South America, the deforestation of which is a major source of carbon emissions. Scientists suggest that up to two-thirds of arable land globally is used to feed factory-farmed pigs, chickens and cattle, as well as to run biofuel-powered vehicles.[11]

Growing feed for factory-farmed animals causes substantial emissions of nitrous oxide through the use of fertilisers. Nitrous oxide is the most aggressive greenhouse gas; 300 times more potent than carbon dioxide, it also depletes the ozone layer.[12]

According to scientists, global nitrous oxide emissions have increased by 30 per cent over the last forty years[13] – a period that has seen major growth in the factory farming of pigs and chickens.

In the US, agriculture is responsible for nearly four-fifths of nitrous oxide emissions, with nitrogen fertilisers the prime cause.[14]

World leaders gathered in Paris in December 2015 to strike a deal to limit global warming to within the two-degree temperature rise deemed by scientists to be the 'safe' maximum level. Even at this level, scientists had already concluded that a third of all land-based plant and animal species could face extinction.[15]

For decades, polar bears have been iconic in campaigns to curb climate change, yet still their numbers dwindle. They sit on top of the world on rapidly shrinking ice packs – as the Earth warms up, they have nowhere left to go. There are currently approximately 26,000 of them worldwide,[16] but scientists estimate that by the end of the century, polar bears will be all but extinct.[17]

Numbers of pigs and poultry, on the other hand, their population running into tens of billions, continue to increase. And, as the burden of cropland needed to feed them expands, emissions of carbon into the atmosphere rise too. As the consequences of global warming become ever more obvious, normally deadpan scientists are making statements that are progressively more alarming. One such assessment predicts a future 'far more dangerous' than is currently believed: 'The scale of the threats to the biosphere and all its life forms – including humanity – is in fact so great that it is difficult to grasp for even well-informed experts.'[18]

With the clock ticking fast, polar bears are increasingly seen as a bellwether for the well-being of our planet. In deciding what kind of future we create for polar bears, pigs and people, the question of how our food is produced could well be a defining one.

LEAVING LINES IN THE SKY

Daybreak and a deepening orange glow lights the path of a woman and child returning to their village with buckets of water carried above their heads. A cockerel crows. A man is busy sweeping the yard with a makeshift broom by the light of a head torch. Soon, sunrise will reveal a flat, sparsely wooded plain. Small settlements dot the landscape along with patches of reddish bare earth. Low mountains rise on the horizon. In the village, children dance barefoot in the dust while parents hang out their washing. A handful of goats roam freely and a black hen leads a scattering of tiny chicks as they forage.

As she returns from the well with her young son France, Anita Chitaya and her husband Christopher begin the challenge of another day. Anita's hair is pulled back and neatly braided, and she wears a cross and rosary beads. She wants the best for her son. She wishes that, for him, the only limit would be the sky itself. But she's also realistic. France dreams of being a pilot. He sees planes passing, leaving lines in the sky. His mother points out that the lines are fumes that contribute to climate change. She recalls her son's reply: 'You know what, Mum, I will be a pilot over there, in the land where they pollute, not here.' She fears for his future because of climate change. 'What they do over there still affects us here', she says.

'Here' is Bwabwa in northern Malawi, a landlocked country in south-east Africa. Known as the 'warm heart of Africa' due to the friendliness of its people, Malawi is among the world's least developed countries and its 20 million people largely depend on agriculture. Recent times have seen the country hit by extreme weather: record flooding in 2015 affected more than a million people and threatened disease,[19] and in 2016 a crippling drought was followed by a crop-devouring pest outbreak.[20] Fears over climate change are growing.

Anita and Christopher farm one acre of land, growing cassava, maize, beans, pigeon peas and pumpkins. They combine crops to improve the soil. The beans return fertility, while the maize produces cobs of corn that provide the family with their staple food. The pumpkin foliage shades the ground, suppressing weeds. She sees the plants as working together. The mix of crops means that at least some will be harvested, whatever the weather. The cobs from the maize go to the pot, while the stalks feed the pigs.

Anita's friend, Esther Lupafya, a nurse at the local clinic, co-founded the local community group, Soil, Food and Healthy Communities, to fight local hunger. The two friends work together to help bring abundant food from dead soil. Yet they have a growing sense that time is against them. The local Rukuru River runs dry more often now. Where once water gushed, it now resembles a beach. Mothers dig for water while their children play. Though they do what they can, things are getting worse.

'We have learned that despite our efforts, climate change is going to continue to get worse because of what they're doing in places like America', Anita told the film-maker and academic Raj Patel in his documentary, *The Ants and the Grasshopper*. Patel arranged for Anita and Esther to visit America, so they could tell their story of how actions there can have profound effects on people thousands of miles away. They flew out to talk to farmers and community leaders in order to find out, in Anita's words, 'why Americans aren't taking climate change seriously'. The visit happened during the Trump administration, when the USA withdrew from the Paris Climate Agreement.

The two women arrived in the States to find a green and 'beautiful' land. At their first stop, an organic dairy farm in Wisconsin, they were astounded to see just how big everything was: the 700-acre farm, the buildings, the tractors. 'It's a different world', said Jim Goodman, farmer and president of the National Family Farm Coalition, and Anita and Esther agreed.

When Anita was told that the tractor had cost $160,000, she laughed in disbelief. It would take five lifetimes in Malawi to afford that. Goodman was keen to explain that the cost of such big equipment was also a stretch in America. 'Most farmers, in order to run as much land as they do, they need that big equipment. And the only way to do it is to borrow. Most farmers are just kind of resigned to the fact that they're going to be in debt as long as they live.'

When Anita brought up the issue of climate change, her hosts put it down to cycles of the weather. Farmers in Wisconsin don't see the polar ice caps melting. Goodman's farmhand Jordan Jamison summed up his feelings: 'I don't see it as an issue – that's my problem.' On another family farm in Iowa, the reaction to the issue was similar. 'For us, it's not a topic of conversation. We see it more as a political agenda', said Tricia Jackson. 'It doesn't feel like a day-to-day thing you think about when you're just trying to manage your farm.'

On another farm, Anita got into conversation about where the crops end up. 'All the food you're growing, the soybeans and the corn, is fed to animals', she reflected to farmer Tyler Franzenburg. He nodded. 'But this would be food for us, in our country', she told him.

Franzenburg described climate change as a 'hot topic' but seemed daunted by the prospect of what to do about it. 'How do we get the entire world on board to try to do something? We can barely get our own local area on board, let alone the United States or the entire world.' Anita gave a look that was somewhere between disbelief and despair. 'While the rich are debating about what to do, poor people have to endure', she said.

It was a tiring journey, mind-blowing and discouraging in equal measure. The bright lights of America felt like a dream, but Anita and Esther also saw the denial and the divide between rich and poor. Driving past homeless people in Oakland, California, Anita remarked that 'the poor people usually look like me'.

When the two women reached their final destination in Washington DC, they hoped to meet US government officials and explain how America was affecting Malawi. They toured the impressive buildings that represent the centre of gravity of the world's most powerful country, but no one would talk to them. Senator Jeff Merkley from Oregon invited them to speak to his team and, although he didn't meet them in person, eighteen months later co-sponsored the Green New Deal, a package of measures to address climate change and economic inequality.

Summing up her time in America, Anita told Patel, 'This country has everything, and yet they take it for granted. Many are still in denial. But as human beings, they should know that nothing lasts forever. As for climate change, many Americans don't fully understand what it means that we in Malawi live on the same planet as them.'

CLIMATE-FRIENDLY FARMING

Prominent on the top of a bare pine tree overlooking endless grasslands, the regal outline of a bald eagle crossed the sky. Her white domed head was adorned with piercing eyes that surveyed the pasture speckled with grazing cattle and roaming chickens. Early-morning mist hovered over the nearby lake like the steam from a recently boiled kettle. She checked herself for a moment before launching, her wings unfolding forward before rowing back with a powerful sweep. She drifted low across the sky, her muscular reflection caught in still, clear waters. Picture-perfect. The moving bird now mirrored in the water. She flew over a bankside boathouse and my camera shutter clicked. With the sun rising, I watched as the eagle passed above me before heading toward distant pastures in search of a meal.

I had come to a farm that to me is as iconic as the bald eagle: White Oak Pastures in Bluffton, Georgia, renowned for its pioneering efforts to produce sustainable food. The farm is the brainchild of Will Harris, a former factory farmer who is now a leading light among the growing number of farmers in America who see farming as a solution to the world's challenges, including climate change.

This is a place where cattle roam rich pastures alongside a multitude of other species. To Harris's colleagues in agriculture, he's a renegade and an inspiration. To consumers, he's producing grass-fed, non-GMO meat without growth hormones. And to his local community, he's Bluffton's biggest employer, with 137 employees in a town of a only hundred residents.

The sandy soil is not good cropland, but it nevertheless used to be heavily cropped with chemical pesticides and fertilisers, with cattle removed from the land and confined to feedlots. Yet since Harris saw the light, the main crop in Bluffton has been grass. I pulled up outside the town's courthouse and was greeted by a big handshake from a big character. Wearing his trademark white Stetson, Harris filled me in on what had been happening in these parts. The courthouse was now his office, though it still had all the legal trappings: judge's seat, witness box and an American flag in the corner. 'That rotten building over there by those trees, that was the jail', he told me. Inside his office, skulls of bulls, rams and longhorn sheep stood on side tables and hung on walls. I looked into a fish tank and a tiny alligator, all head and bulging yellow eyes, peered back at me. 'I rescued that young man last fall, in a drought', Harris said. 'I'm waiting for some rain, to set him loose.' His desk was full of new plans for the farm. Bluffton may be remote, but for those interested in sustainable food, this place has become a real draw.

A fourth-generation cattleman, Harris farms the land his great-grandfather settled upon in 1866. Graduating from Georgia's agriculture school in 1976, his upbringing centred on industrial farming. His father was focused on getting the most out of the land through monocultures, be they fields of the same crop or cattle crowded together, raised with hormones and antibiotics and fed on corn and soya. The men in the Harris family were alpha males with the philosophy 'more is better'. If the instructions on the pesticide said a pint, they'd put a quart. There was no holding back.

They were factory farmers, and good at it, so why change? 'Nature abhors a monoculture', Harris told me while explaining his change of heart.

In the 1990s, Harris embarked on a journey that would see him become one of America's trailblazers for regenerative farming. Intensification had taken its toll on the local wildlife, water and soil, so he took the farm in a completely new direction. 'And now, the further we move away from it, the more I enjoy it. And it's become my passion.'

Far from being confined, his cattle now roam across wide pastures and feed on grass. His farm comprises a rotating mix of 2,000 cattle, 1,200 sheep, 750 goats, 1,000 pigs, 60,000 chickens for meat, 10,000 laying hens and 1,000 turkeys. As if that wasn't enough, it also has 1,500 ducks, geese and guineafowl, fifteen guard dogs, nine horses, and 'countless microbes in the soil'.

The animals are constantly on the move around the 2,500 acres of warm-season perennial pasture and forest. The herds and flocks follow a 'Serengeti rotational model', with large ruminants followed by small ruminants and then birds. Harris sees it as emulating the movements of wildebeest on the ecologically rich grasslands of Africa. It also recalls the wanderings of the bison on the Great Plains.

Here at White Oak Pastures, the herds adhere to the 'law of the second bite'; animals are moved before they can take that 'second bite' of the same plant, allowing the land to rest and the vegetation to recover. Cattle are adept at grazing on long grass, while sheep prefer it shorter. And the chickens like short grass and the insects from the cowpats. 'My bald eagle population has exploded.' Harris told me, explaining how he sees his pasture-raised chicken as supplementary food for what he estimates to be a hundred eagles in the area.

Harris sees this approach as having benefits that range from animal welfare to soil restoration and tackling climate change. Putting animals back on the land in the right way in mixed, rotational farming can provide the very best conditions for animal well-being. 'What we are trying to do is to create a system that allows our animals to follow their instinctive behaviour. It's a silver bullet for animal welfare,' Harris said.

On climate change, there are two main ways to stop it getting worse: by reducing greenhouse gas emissions and by removing carbon dioxide from the atmosphere. The UN FAO estimates that the world's soils could capture or 'sequester' 10 per cent of human-induced emissions,[21] while a study by the Rodale Institute found that a fifth of human-induced greenhouse gas emissions could be offset if half the world's cropland took a regenerative approach.[22] Regenerative farming reintegrates farm animals in a way that fertilises the soil naturally, looking after the microbes and other biodiversity underground while helping them to take carbon out of the atmosphere.

Harris has been instrumental in setting up a new body, the Regenerative Organic Alliance, with the strapline 'Farm like the world depends on it'. By adopting regenerative organic practices around the world, the alliance believes we can create long-term solutions to some of the biggest issues of our time.[23]

Cows are a mainstay on Harris's farm, even though they have come to be seen as bad news in the climate debate for their emission of methane. However, not all beef is the same in climate terms; it depends how the cows are kept. The most intensive is the lifetime feedlot, where cattle never get to roam on grass. Then there's 'conventional' rearing, where cattle graze for much of their life but spend their last few months in confinement. And then there are cattle kept on pasture for their whole life, like Harris's.

Harris has figured out how to produce high-welfare, carbon-neutral beef, and the secret lies in that regenerative approach. He has secured the backing of the American multinational food giant, General Mills – behind such well-known brands as Cheerios, Nature Valley and Häagen-Dazs ice cream – and brought in a company called Quantis to compare the carbon footprint of White Oak Pastures with that of intensively raised beef. The findings were a revelation. Compared to conventional US beef production, White Oak Pastures' carbon footprint per kilogram of beef was found to be 111 per cent lower.[24]

Harris's system is effective at capturing soil carbon across the board; his operation offsets about 85 per cent of its overall emissions. While questions remain about how long the carbon stays in the soil, the results of the study suggest that regenerative cattle rearing has a 'net positive effect on climate'.[25] The findings chime with those of other scientists who have found that methane emissions from grazing cattle can be more than offset by carbon sequestration in the grasses and soil they graze on.[26]

Regenerative farming is based on maximising the efficiency of our use of 'free' resources – like grass and water – while reducing greenhouse gas emissions per hectare of land. The best way to produce healthy meat with the fewest resources is to use permanent pasture or keep animals on the grassland rotation of a mixed farm. In the latter routine, soils are rested from the relentless demands of arable cropping by turning them into grazing land for a few years.

Harris argues that a regenerative system like his might, in the long term, be the only way to feed the world. 'I was in industrial agriculture for the first forty years of my life, and I still associate with those people ... That [industrial] system may feed more people in the short run. But in the longer view, I would argue that my system will feed more people.'

Regenerative agriculture, coupled with a reduction in the number of animals reared for food worldwide, promises an environmental win-win. Fewer animals, reared in ways that restore our contract with the land and bring back bald eagles and other wildlife, holds the key to true sustainability, and Will Harris is just one of the people holding that key firmly in his hands.

REBEL WITH A CARROT

Bright pink and pulling a trailer full of vegetables, a flamboyant tractor chugged its way to London intent on being a showstopper. Draped in banners, powered by biodiesel and followed by a convoy of renegade farmers, this cavalcade was bound for Westminster. But this was not an uprising about fuel prices, subsidies or countryside pursuits. Rather, rural England was heading into the heart of the capital to join climate protests by Extinction Rebellion (XR), part of a two-week rally in sixty major cities worldwide, including Paris, New York and Delhi.

The protest seemed unstoppable – that is, until the pink tractor broke down. Undeterred, another rural chariot, a green tractor this time, was wheeled out to lead the charge. It took the convoy three days to get to London; though they failed to reach Parliament, they made it to Chelsea, where the police intervened. The media had a field day, reclaiming the phrase 'Chelsea tractor', previously reserved for urban four-by-fours.

'We had flags, we had big boxes of veg, we had it all going on. It was brilliant', said Dagan James, the man behind the wheel of that purring pink tractor. Inspired by earlier Extinction Rebellion protests and Greta Thunberg's youth strikes aimed at persuading governments to respond to the climate emergency, James saw how farmers needed to be involved. He founded a spin-off group, XR Farmers, made up of sheep rearers, vegetable growers and arable farmers, part of what he described as a 'movement of movements'.

'The science is clear – we're running out of time', he said. And it's true that without action, global heating is likely to turn the Earth into a hothouse. An increase of two degrees would see 99 per cent of the world's coral reefs disappear. The Greenland ice sheet would be lost, leading to sea-level rise of up to seven metres, swamping coastal towns, cities and low-lying farmland. It would also have severe consequences for major economies including the UK and Europe, both financially and in our ability to grow food.[27]

How big a contributory factor is food to the problem? Well, scientists estimate that if we carry on as we are, food will be responsible for more than half the emissions it would take to reach two degrees of global heating,[28] with the majority of these coming from animal farming.[29] From talking with James, I got the impression that farmers were increasingly seeing farming not just as part of the climate problem, but as part of putting things right. 'We're not blaming any individual farmers here at all', he told me. 'Yes, of course there are problems with farming, but we need a change in the system to enable farming to adapt and respond and become part of the solution.'

In his view, looking after the soil and reconnecting farm animals to it as a big part of the solution. An advocate of regenerative farming, I could sense his frustration at the damage being done to the environment: he referred to the 'awful mismanagement of our life-support system'.

In leading farmers in climate protests, James had broken the mould, but I didn't expect to hear him talk so clearly about the role of diet. He thinks we need to change what we eat: 'Much less meat. Less animal products ... We have to attach more respect to it; we have to eat less of it.'

Most animal products cause far more greenhouse gas emissions per unit of protein than plant products such as grains, vegetables and pulses.[30] A recent Chatham House study points out that global temperature rise is unlikely to stay below two degrees Celsius if we don't reduce our consumption of meat and dairy.[31] And a 2020 report by the UN FAO, a body not known for its radicalism, concluded that animal products are responsible for more than three-quarters of food-related greenhouse gas emissions. Moving to diets that range from eating less meat – known as flexitarianism – to being vegan would reduce the costs arising from food-related emissions by between 41 and 74 per cent.[32]

As well as needing to reduce overall meat consumption, James argues that we should remember what it means to obtain meat: killing sentient animals. He had invited me to his farm, Broughton Water Buffalo in Hampshire, to see his pasture-raised water buffalo and to get up close and personal with *that* pink tractor, restored to working order. As we approached, about a hundred water buffalo crowded curiously at the fence line. They had big endearing faces, dog-like eyes and a trusting, almost questioning look. Bulky, broad heads were framed by bony, round horns. They examined us closely, while making muffled snorts.

'I don't really like killing them', James confessed. 'Ultimately, we may not have farm animals. Ultimately, we probably won't eat meat, down the line. But I would say, in this adaptation phase, this time of massive transition, I think there is a place for livestock.'

For now, he sees animals as part of the solution; fewer of them, integrated into rotational mixed farming. 'A key thing about it is getting animals out into fields as part of rotations in arable farming systems. And if you do that, you're producing a grass-fed healthy product that is delivering lots of other services: biodiversity, soil organic matter, soil health, nutrient cycling … and employment', he said.

And the end game? 'A completely regenerative farming model across the world.'

THE NATURE OF HOPE

My time with Dagan James, Will Harris and other pioneers got me thinking about hope, and how it has to be predicated on action if it is to be anything other than false hope.

My thoughts were piqued when I was on a *Question Time*-style panel at the British Birdfair where a journalist posed the question: 'When it comes to climate change, what evidence is there that gives real reason for hope?' I was in great company. My fellow panellists included Chris Packham, Deborah Meaden, chair of Natural England Tony Juniper, author and rewilding pioneer Isabella Tree, and Carrie Symonds, wife of prime minister, Boris Johnson. As we discussed the topic, the main point that emerged was the need for hope and positivity if we are to motivate people to do something about climate change.

But current efforts on tackling climate change are providing little cause for optimism. So much more needs to be done, in so little time.

It made me think; hope comes from the belief that things can and will change for the better, but is it really hope that changes things, or is the big driving force of change anger rather than hope?

After all, in many cases, social change like action on climate chaos or animal cruelty is motivated not by hope, but by indignation, a sense of frustration and injustice. Looking at my own situation, I first got involved in animal protection issues not through a sense of hope that things would improve, but out of anger at the treatment of animals in the meat industry, at factory farms, during long-distance transport and at the slaughterhouse. I remember spending sleepless nights just thinking about the animals' plight. I had a burning feeling that things were wrong, and I wanted to be part of bringing about a better way.

Hope, for me, came from finding ways to get involved and from making my voice heard. I joined a local animal protection group, got involved with organisations like Compassion in World Farming and started to read more about the subject. I saw books as weapons – I armed myself with the facts and came out fighting for change. It wasn't only animal issues; in my teens, I was also active on issues like global peace, pollution and the miners' strike.

I didn't want to be patronised with soothing stories about how things weren't as bad as I feared. I wanted realism and honesty, but most of all, I wanted to help bring about change. And to me, a big part of feeling hopeful came from seeing that I wasn't the only person who felt like this. That I wasn't alone in wanting issues dealt with rather than brushed under the carpet.

For me then, hope starts with believing that something must and *will* change. That if *they* don't take action to make it change, *we* will. In never taking no for an answer.

As the teenage climate activist Greta Thunberg said from the TED Talk stage, 'We do need hope, of course we do. But the one thing we need more than hope is action. Once we start to act, hope is everywhere. So instead of looking for hope, look for action. Then, and only then, hope will come.'[33]

In a world where we have a matter of years to solve climate change, where swathes of our food are threatened by the decline of pollinating insects, where wild fish stocks could be gone within a generation and where our soils might disappear within decades, hope for hope's sake isn't enough. As Extinction Rebellion would say, this is not a drill.

Hope without the prospect of action is false hope, which should make us very angry indeed. To my mind, the most meaningful hope comes from people taking action to change things themselves. And I've come to learn that we can all act three times a day, in the meals we eat. Choosing to eat more plants and less but better meat, milk and eggs can help unlock a healthier diet, reduce animal cruelty and head off the worst excesses of climate change. In this way, we create a hope that is rooted in realism and a tangible sense of progress.

8

WILL INSECTS SAVE US?

BRINGING BACK A MILLION BEES

Early summer, and a bee buzzed around a tan-coloured sow as she wallowed in her drinking trough. Lying enjoying the cooling water, she watched as the bee hovered by. Spotting me, she lifted herself up to her trotters and shook herself, spraying water in all directions. Nonchalantly, she wandered off. My gaze switched to the broad strip of colourful flowers that stretched beside the outdoor pig arcs. Bunches of phacelia blooms burst from waist-high stalks, attracting passing bumblebees. Laden with pollen, the furry insects were flying busily from flower to flower, creating a buzz of excitement.

Twenty years ago, the Hayward family were leading the charge in outdoor pig rearing. Back then, they were establishing their now internationally successful Dingley Dell Pork brand. Mark had a dark ponytail; I was not long out of a mullet. Now we both sport greying close-cropped hair befitting of our 'been there, done that' time of life. Mark's family has owned this farm near Woodbridge in Suffolk for a century. It used to be a 'super-mixed' farm, with dairy and beef cattle as well as pigs.

When I visited two decades ago, I remember seeing a turtle dove fly over the pig pens. Across the country, they have since suffered severe declines, but here at Ashmoor Hall Farm they still have two pairs visiting. They have barn owls too, another once-common farmland species that's rarer now in the UK.

Twenty years later, it wasn't pigs that brought me back to this pioneering pig farm in Suffolk but plans to save endangered pollinators.

Farming brothers Mark and Paul had come up with an idea to bring back a million bees. They'd measured it, tested it and figured out how to make it scalable. And now they were eager to share their idea. Their plan involved planting flowering plants for bees as part of a mixed rotational system, with crops and animals moving around the farm in succession. Large patches of nectar-rich plants follow the Haywards' pigs, followed by maize, oats, wheat, barley and oilseed rape.

For the bees, the spikey blue phacelia flowers were a firm favourite. A count of bees at Ashmoor Hall revealed that a single square metre of phacelia would support up to twelve bumblebees. They planted 338,000 square metres of flowers – equivalent to eighty-three football pitches – in blocks around the farm, enough to attract over a million bees. Other plants included sainfoin, bird's-foot trefoil, alsike clover, musk mallow and vetch.

'Why would you want intensive pigs when you could be doing this?' Paul asked. A sow fixed me with a stare as she lay in her straw-lined compartment, her ears twitching and lashes fluttering. Her piglets nosed and chased each other. Some were ginger with black spots, while others were a more familiar pink. Their intense curiosity shone through. One moment, they'd be standing and staring intently; the next, they were off with a grunt, running like puppies with ears flapping in unison as they bounded forward and back. These were all RSPCA-assured and 'outdoor bred and reared'.

Mark and Paul wanted to develop the farm and thought outdoor pig farming had to up its game to become more environmentally friendly. With 900 sows producing 22,000 fattening pigs a year, they began to look for ways to keep the land healthy. Now, when the pigs are rotated around the farm, the land is rested to avoid parasites and disease. And seeding it with phacelia and vetch helps to hold nitrogen in the soil, which cuts pollution.

'We thought, rather than planting grass, let's plant something that feeds things', Paul told me. 'We could be feeding a load of bees, a load of insects and butterflies with the land, rather than just having some rye grass that binds the soil but doesn't actually feed anything. Little changes mean we can do so much more.'

'And if you focus on bees, you feed the other insects too', Mark added.

Some 40 per cent of insect species are in danger of extinction. Bees are essential for the pollination of a third of all our food and about three-quarters of all crops worldwide, yet they are struggling. To put an economic figure on it, between $235 billion and $577 billion worth of annual global food production relies on the contribution of pollinators.[1]

What the Haywards have been doing in Suffolk might provide a template for saving bees elsewhere. I asked Paul what his advice to other farmers might be. 'Become a little untidier … Let nature have a chance', he said. 'You can help nature a bit more by taking out those corners of fields that everyone knows aren't farmable. Do something with them: put in a bee mix, put in a mixture for wild birds to feed on over winter. Let your hedges grow a little bit bigger. Put a margin in.'

The Haywards leave 15 per cent of their farm for nature, though they told me that, so far, their neighbours haven't followed suit. Instead, they continue to farm intensively, ploughing right up to the hedges and leaving no room for wildflowers that could be a lifeline for bees, butterflies and birds.

For the Haywards, farming is about making a living today and putting something back for tomorrow. But can a more ecological approach to farming really pay? Paul is confident that it can. 'It's really quite simple: buy a slightly smaller tractor. Cut your cloth accordingly, so you can be as profitable on a smaller area as you could be on a slightly bigger area with a bigger overhead.'

As for the future, they believe that meat production has to align with sustainability. 'I would like to see the entire farming industry realise that they are the solution to the wildlife in this country. And with some clever ways of funding wildlife projects on every farm in this country, we could do a massive amount', Paul said.

Paul and Mark each have three children and hope they will take over the farm one day. For Paul, the project is about their future.

'That's what I want my grandchildren to see ... I want them to be able to walk through a meadow and hear the birds, see the dragonflies, see the diversity of plants and therefore animals ... I want them to have what I had.'

FARMING INSECTS

Cameras popping, King Willem-Alexander of the Netherlands strode purposefully into a new factory in the city of Bergen Op Zoom. Having donned a protective apron and blue gloves, his hands were soon plunging into rotting vegetables teeming with maggots. Behind tinted glass, black specks with twitching antennae clung to the illuminated surface. These were black soldier flies, and in the wild, their larvae play an essential role in decomposition, helping to break down dead stuff and return nutrients to the soil. And they were part of a demonstration of what some see as the food and animal feed of the future: insects.

The Dutch king was present for the grand opening of one of the world's largest insect farms. The company behind the operation, Protix, was keen to reassure journalists that the king was at no risk of harm. 'He was unafraid', CEO Kees Aarts told Reuters. As for the insects, Aarts described 'The sensation of the energy that comes from the growth of the insect' as 'beautiful'.[2]

The bug production business was hotting up amid claims that, other than going vegan, eating insects was the only way to save the planet. Yet insects are already part of the diet of about 2 billion people, especially in tropical regions with high levels of biodiversity.[3] Beetles, butterflies, moths, crickets, grasshoppers, wasps and ants are among the most popular insects consumed,[4] with most of them harvested from the wild[5] before being steamed, roasted, smoked, fried or put in a stew.

To many of us in the West, eating insects is not that appealing. In Britain, maggots are used as fishing bait and freeze-dried mealworms are used as garden birdfeed. Few people see them as food. But in order to overcome Western sensibilities, farmed insects are being powdered or turned into flour.[6] Insect burgers made from buffalo worms and organic soya have popped up in Brussels restaurants and supermarkets. Baris Özel, managing director of Bugfoundation, the company behind the innovation, told *FoodNavigator*, 'Our vision is to change the eating habits of a whole continent in a sustainable way.'[7]

As the idea of eating insects has taken off, so too has the industry around farming them. In North America and Europe, the creatures are raised under controlled conditions before being killed by freezing or shredding.[8] They are then freeze-dried, packed or pulverised, before being either eaten as they are or added to burgers, bakery products or snack bars.

EFFICIENT?

Insects are, without doubt, less resource-intensive than traditional farmed animals, which is a reflection of how inefficient cattle, chickens and pigs are at converting grain into animal protein. However, intensively farmed insects are commonly reared on cereals and soya, too. When scientists looked at the amount of feed needed to farm crickets compared with carp, chicken, pork and beef, the 'feed conversion ratio' (the amount of feed in compared with the amount of food out) for insects was found to be marginally better than carp, twice as efficient as chicken and nearly twenty times better than beef.[9] However, it's important to realise that crickets and other bugs fed on crops are not a 'free lunch'; it is far better to use croplands to feed people directly than to feed farmed insects.

Black soldier flies and Argentinian cockroaches are among the most efficient insect species, with food conversion ratios of between 1.4 and 2.7 to one, but even they eat more food than they produce.[10] So with farming insects as opposed to foraging for them in the wild, it's the same old story – a system that uses more food than it actually produces. Added to which, the consumption of insects is not without safety concerns; they have been shown to accumulate hazardous chemical pesticides, heavy metals, pathogens and allergens.[11]

WHY?

This brings us to the question: what's the point? Protagonists claim that insects can help feed a growing population, an argument based on the false premise that we can't grow enough food for a projected global population of 10 billion; yet we already produce enough to feed twice the current human population. The main reason for the 'not enough' illusion is the amount lost due to food waste and the quantity of cereals and soya fed to livestock, whether chickens, cows or crickets.

Just as factory farming is wasteful and inefficient, so too is the large-scale production of crop-fed insects. The rationale for industrial insect farming becomes even more tenuous when the purpose is to feed livestock. Perfectly good crops are fed to insects, which are then fed to pigs, chickens and fish. As Phil Brooke, welfare and education manager at Compassion in World Farming, wrote, 'factory-farmed insects producing feed for factory-farmed pigs and poultry ... would create suffering on a massive scale for the pigs and chickens and, potentially, on an astronomical scale for the insects. All of this to waste grain which could have fed the hungry.'[12]

CHEMICALS

The industrial production of animals – insects or otherwise – relies on the industrial production of cereals, which usually means spraying the countryside with insecticides. So to produce one set of insects we wipe out another, impoverishing an ecosystem and leaving the songbirds and other creatures that rely on wild insects for food to go hungry in the process. There is an irony, then, in our turning to insects to feed us while also waging war on them in the countryside. Chemical sprays kill insect communities indiscriminately and often for several years, undermining the sustainability of our food.

However, scientists are increasingly showing that reducing or eliminating the use of pesticides is not only right for tomorrow but that it needn't limit food production today. A recent French study found that low pesticide use wasn't necessarily a barrier to high yield and profitability.[13]

WELFARE

The question of whether farming insects is ethical from an animal welfare perspective is also important, and the obvious caveat is that insects often like conditions that we don't. I remember a colleague telling me about the awful smell of a maggot farm she visited. 'Perfect conditions for maggots, though', I couldn't help but think. When considering the welfare of insects, the key thing is that we keep them in a way that allows them to fulfil their natural behaviours, however repulsive we might find them.

Although scientific understanding of the inner workings of insects is limited, there is evidence that they have the capacity to suffer. They have receptors that sense heat or injury, while honeybees have been shown to have the capacity for optimism and pessimism. Ants teach each other where to go for food, with teachers having less patience with slow learners. When Matabele ants are injured during raids on termite colonies, they are carried home and their wounds licked to prevent infection. But the most intriguing behaviour by ants is their response to the 'mirror test'. If a blue dot is painted on an ant's head, they see it in a mirror and will try to clean it off. Such a response is regarded by scientists as proof of self-recognition; that the ants can see themselves.[14] It offers further evidence that insects should be treated with compassion.

THE COUNTRYSIDE

The impending rural farmageddon is leading to what some have described as 'insectageddon'. As farming intensifies, humanity's ecological life-support system is being reduced to a pale shadow of itself, existing within a toxic environment where even nature reserves can't help. Scientists in Germany found that the biomass of insects in the country's nature reserves had declined by more than three-quarters in just twenty-seven years,[15] with agricultural intensification the likely cause.

An international team of scientists reported that up to half a million insect species had become extinct in recent times, with half a million more likely to in the coming decades. They blame loss of habitat and the use of harmful agri-substances, as well as climate change and the spread of invasive species.[16] Some scientists believe that imminent decline could cause the extinction of 40 per cent of the world's insect species.[17]

MAKING PEACE

Without insects, the world would be a very different place. In the countryside, they are essential for pollination and the decomposition of organic matter in the soil, as well as feeding the web of life on which we all depend. Without bees, our local supermarket shelves would look very different. There would be no tomatoes, chilli peppers, courgettes, blueberries, raspberries, runner beans or cucumbers – the list is endless.

On sustainability, the key question is whether intensification leads us to squander crops to feed insects, or to farm insects for animal feed. On my analysis, I find it hard to justify large-scale insect farming. If we are looking for efficient ways of producing protein, what about cultured meat or plants? And as for ethics, it's hard to know how much thought has been given to insect welfare.

Making peace with insects is a prerequisite if we want a sustainable future. We must use beneficial insects to tackle those that are considered 'pests', and use the secrets of the ecosystem to keep things in balance. There is increasing recognition that farming has to change, and a growing band of farmers are already making space for insects among their harvests. To me, it is their efforts, not the latest insect farm, that should be receiving royal attention, for it is they who are ensuring a countryside with insects that will be far better able to produce food fit for a king.

9

WHEN OCEANS RUN DRY

OCTOPUS FARMING – MEETING THE 'INTELLIGENT ALIEN'

Bulbous eyes peered prominently from a mass of swaying flesh, as the pink creature stretched sucker-covered tentacles around string dangling at the bottom of an otherwise empty tank. Another octopus clung to the sides of a blue plastic pool. Conditions on this model farm appeared more akin to a laboratory. As the promotional film continued, the narrator marvelled as she cradled the octopus's head like a baby. 'Look at that creature', she said invitingly. 'You are amazing.'[1]

According to the scuba-diving philosopher of science Peter Godfrey-Smith, the octopus is 'the closest we will come to meeting an intelligent alien'. In his book *Other Minds*, he describes them as 'an island of mental complexity in the sea of invertebrate animals'. The brain of a common octopus has 500 million neurons, making it as smart as a dog or a three-year-old child. But unlike vertebrate animals, an octopus's neurons are arranged throughout the entire body. These amazing creatures are 'suffused with nervousness' – including the arms, which act as 'agents of their own' – and sense by taste as much as touch. According to Godfrey-Smith, the usual body/brain divide does not apply to them.[2]

By any measure, octopuses are truly remarkable: they have eight legs, three hearts and blue-green blood. Masters of camouflage, their skin is embedded with cells that sense light, giving them an array of tricks for thwarting enemies. They can match the colours and textures of their surroundings, enabling them to become near invisible in plain sight. They can escape at speed by shooting forward with jet propulsion. They can squirt ink to hide themselves and dull the senses of an attacker. And if they lose an arm, they can grow it back.[3]

As the well-known ethologist Dr Jane Goodall observes, octopuses are 'highly intelligent' and able to do extraordinary things. Wild octopuses have been known to carry coconut shells with two tentacles and walk across the ocean floor with the other six, whereupon 'they put down half the shell and ooze their body into it because they are very soft-bodied and can get into tiny spaces. Then they reach out and put the other half over them. They've made a house!'[4]

PROBLEM-SOLVERS

Members of the cephalopod family of invertebrates, octopuses are curious, explorative problem-solvers with long memories. One study found that they remembered how to open a screw-top jar for at least five months.[5] They can also show a sense of craftiness – squirting water at researchers they don't like. One celebrated aquarium-kept octopus became renowned when staff noticed that fish from a neighbouring tank had gone missing overnight. CCTV revealed that the octopus was lifting the lid of his tank, slithering over to grab the fish and then crawling back, before returning to the tank and putting the lid back on as if nothing had happened.[6]

Octopus has been fished from the oceans for two millennia,[7] but in recent decades, the numbers caught have almost doubled, from 180,000 tons in 1980 to more than 350,000 tons in 2014.[8] Asia accounts for two-thirds of the global octopus catch, with China alone accounting for more than one-third. Demand is also growing in the United States and Australia. The main importing countries are Japan, Korea and the northern Mediterranean countries of Spain, Greece, Portugal and Italy.[9]

As stocks of fish around the world continue to decline, demand for octopus is increasing, yet their numbers are coming under strain. Overfishing, combined with this growing demand, is driving prices up, leading to burgeoning interest in the farming of octopus.[10]

The potential of octopus farming is being explored around the world. In Portugal and Greece, Nireus Aquaculture has funded research, and octopus ranching is being tried in Italy and Australia. A farm in the Yucatán Peninsula of Mexico has been reported as successfully farming the Maya octopus, and attempts to farm octopus are under way elsewhere in Latin America. In China, eight different species of octopus are being experimentally farmed, while the Japanese seafood company Nissui has reported successfully hatching octopus eggs in captivity.[11] In Australia, octopus has gone from accidental catch to gourmet food so quickly that the fishing industry has been unable to keep up with demand, further throwing a spotlight on octopus as a potential candidate for aquaculture.

PIONEERS

Ross and Craig Cammilleri of the Fremantle Octopus Company were keen to explore how they might raise octopuses in captivity. The brothers wanted to move from fishing for the creatures to 'ranching', where wild-caught juveniles are grown in land-based tanks or offshore cages. However, they quickly found that it was impossible to catch enough juveniles to make large-scale 'ranching' commercially viable, so they approached the Department of Fisheries for Western Australia.

A four-year research project culminated in the establishment of a fifteen-tank model octopus farm north of Perth. An industry was soon formed that involved bringing back young octopuses and growing them in captivity, thus utilising the two-thirds of undersized creatures fished from the oceans. It worked well because tank-reared octopus grow fast, singling them out as prime candidates for farming.

One of the problems for the industry was how to keep these intelligent, inventive creatures confined in a tank. Octopuses developed a habit of hurling themselves out of the tanks, and neither heavy steel mesh nor electric fences were enough to prevent them climbing out of the tanks. Placing woven cloth around the edge of the tank eventually stopped them; their suckers couldn't gain a grip on the porous material.[12]

In the wild, octopuses will go to great lengths to defend their patch, a behaviour that limits the number that can be reared in a given space. In a farmed environment, with lots of animals in a sterile tank, things can quickly become fraught, leading to aggression and cannibalism. Researchers have overcome this problem by providing each octopus with a plastic tube to hide in, though that limits the number of creatures they can cram into a tank.

FARMING INTENSIVELY

At the model farm in Western Australia, researchers found that aggression was reduced when they kept individuals of the same size in a tank, with no need for plastic hides. This made cleaning the tanks easier and meant they could keep more than triple the number of octopuses in each tank. The lead researcher Dr Sagiv Kolkovski was quoted as saying, 'We added so many to the tanks that we had to install flat PVC sheets so they would have more surface to attach to, as the tank walls were completely occupied.'[13]

Intensive octopus farming was becoming a real prospect, but there was still a huge problem to overcome: no one knew how to rear octopus in captivity past the larval stage. Under research conditions, females had laid their eggs in captivity, which had gone on to hatch successfully; researchers in Australia subsequently claimed that octopus larvae had, for the first time in the world, been 'born without a mother'.[14] However, the problem of how to provide the right nutrition and environmental conditions to grow larvae remained.

The breakthrough came in 2019 when the Spanish Nueva Pescanova Group announced that common octopuses born in confinement had not only been reared to adulthood, but one of them had gone on to produce eggs. It was sufficient progress for the company to announce that it expected to be selling farmed octopus by 2023.[15] The company went on to declare that its pioneering work was in 'response to the high international demand' in recent years, which has 'caused a growing scarcity of wild octopus and, therefore, a sustainability problem in the marine environment'.[16]

MISTAKES REPEATED

In the rush to farm these 'intelligent aliens', as Godfrey-Smith describes them, fears have grown that the likely impact on the marine environment is being overlooked. Octopus eat small fish and other marine life; their diet in aquaculture will most likely consist of fishmeal made from the fish that are eaten by bigger fish, birds and marine mammals. So rather than protecting the oceans, the farming of carnivorous species like octopus, as well as salmon and trout, puts yet more pressure on the oceanic environment.

As four scientists, including Peter Godfrey-Smith, wrote in 'The Case Against Octopus Farming': 'Like other carnivorous aquaculture, octopus farming would increase, not alleviate, pressure on wild aquatic animals. Octopuses have a food conversion rate of at least three to one, meaning that the weight of feed necessary to sustain them is about three times the weight of the animal. Given the depleted state of global fisheries and the challenges of providing adequate nutrition to a growing human population, increased farming of carnivorous species such as octopus will act counter to the goal of improving global food security.'[17]

The authors contend that, even if a more sustainable diet for them could be found, taking intelligent creatures with complex behaviours and rearing them in barren conditions would be unethical. It remains to be seen whether intensive octopus farming would also come with environmental impacts such as pollution, disease and the overuse of drugs common in other forms of intensive aquaculture.

What concerns me is whether humans are being 'humane' in this new farming venture, respecting octopuses as creatures with great sensitivity and notable intelligence. On current form, the plight of the farmed octopus makes me hope that we never meet an 'intelligent alien' in space. That way, we can't let ourselves down further.

FISH FARMING – KILLED BY THE CURE

Mid-August 2020 on the west coast of Scotland, and Storm Ellen was rattling windows and battering the coastline, whipping the sea into a frenzy and obscuring normally stunning views of the Island of Arran. Meanwhile, on the east side of the Kintyre Peninsula, a calamity was brewing. An open-sea salmon farm, normally anchored to the seabed, was in motion. A feed barge had come adrift in the wind, its mooring lines cutting across those holding ten circular pens in place. Friction took its toll and the floating farm became dislodged, along with half a million salmon. Buffeted by the storm, four pens crumpled, nets were torn and nearly 50,000 fish escaped to freedom.[18]

The storm caused an embarrassing setback for an industry being touted as Scotland's 'new oil'. With revenues from North Sea oil declining,[19] there was much talk about the nation's 'blue economy', which focused on marine and coastal development. From just a couple of sites about fifty years ago, more than two hundred fish farms now operate in Scotland, producing more than 200,000 tonnes of salmon worth over £1 billion a year.[20] Exports accounted for £618 million in sales revenue in 2019,[21] with industry sights set on doubling in value by 2030.[22]

When salmon escape from farms they have a big impact on their wild cousins, competing for food, infecting them with parasitic sea lice and interbreeding with them in a way that puts their future in jeopardy. Wild salmon catches are in crisis, with government data showing them to have hit their lowest level since records began.[23] Among those expressing concern at the decline is Prince Charles; at the launch of a new angling and conservation initiative, the Missing Salmon Alliance, he warned that 'The very future of a species that has been swimming in our oceans and seas for over six million years will be in jeopardy ... We simply cannot allow this to happen in our lifetime.'[24]

The ramifications of industrial aquaculture are hard to discern, but in the winter of 2020, a diver took to the icy waters off the west coast of Scotland, the Isle of Skye and Shetland, to uncover salmon farming's hidden side. Commissioned by Compassion in World Farming, the investigation found salmon with pink sea lice around their heads and eyes in obvious infestation. These parasites feed on salmon's scales, flesh and tissues, leaving them with painful open wounds and susceptible to disease. Though they occur naturally in the wild, they spread particularly virulently in salmon farms, where fish are crowded together. Some fish showed diseased and swollen gills. One swam in a bit of a daze, with eyes missing and sockets red raw. Some had chunks out of their bodies and gaping wounds, while others had seaweed growing from their injuries, as if they were rotting alive. The 'king of fish' had clearly taken a fall. Investigative footage also showed bloated, spiny lumpfish in a murky pen – they'd been thrown in to nibble the sea lice, but showed signs of being attacked by lice themselves. It was a pitiful sight.

It is little wonder that salmon farms regularly record mortality rates of up to 30 per cent, despite measures to keep the fish alive that include pumping them through hot water to remove sea lice in a contraption known as a thermolicer. Alternatively, the fish may be crowded together and sprayed with water jets, in the hope of dislodging parasites. Some farms use hydrogen peroxide, essentially bathing the fish in bleach. In any other type of animal farming, such high death rates would trigger serious questions; in salmon farming, they are taken as normal.

The conditions in which farmed salmon are kept is a major reason why mortality rates are so high. Packed tightly, with each fish having the equivalent of a single bathtub of water, these natural ocean wanderers are forced to swim in incessant circles, just as frustrated zoo animals pace their enclosures.

And nor are other species treated much better; in fact, life for farmed trout can be even worse. Generally reared in freshwater 'raceways' or earth ponds, they can be kept at densities of sixty kilograms of fish per cubic metre of water – the equivalent of twenty-seven trout being kept in a bathtub.

Fish farming has become big business internationally, and salmon is only part of it. Globally, more seafood is farmed than is caught in the wild, and the production of fish and seafood has quadrupled over the past fifty years. The world's population more than doubled during this period, and the average person now eats almost twice as much seafood as half a century ago.[25] Asia has led the way, accounting for more than four-fifths of fish farming worldwide.[26] Despite a common perception that fish farming takes the pressure off overfished oceans, when the species farmed is carnivorous, as salmon and trout are, the reverse is true. A significant proportion of their diet comes from wild-caught fish ground into fishmeal. A single farmed salmon takes about 350 wild fish to produce.[27] To put it another way, the Scottish industry feeds as much wild-caught fish to its salmon as is eaten by the entire UK population.[28] In terms of taking the pressure off the oceans, it's a case of being killed by the cure.

In thirty years of writing about the fish farming industry, I've found few reasons to suggest it pays anything other than scant regard for the welfare of the animals in its charge.

However, one thing that has changed recently, in Scotland at least, is the outlawing of the shooting of seals around fish farms. Huge quantities of fish on a farm present an irresistible temptation to seals, rather like a well-stocked garden bird table attracts local wildlife. The common response of the salmon industry has been to shoot the seals drawn by the prospect of an easy meal. For years, the Scottish government has been reluctant to do anything that might upset a big-money industry, but in February 2021 they introduced a ban on fish farmers shooting seals. This action wasn't brought about by concern for animal welfare, but a desire to save annual salmon exports of nearly £200 million, thanks to new US legislation to protect marine life. Fisheries that kill or injure marine mammals are now banned from exporting their produce to the US. While the ban is cause for celebration, keeping track of what goes on at remote coastal locations is not easy. It is therefore hard to say to what extent the ban will be implemented. An industry headline – 'The predator that's killing 500,000 Scottish farmed

salmon a year' – hinted that there may be trouble ahead.[29] While
Scotland's seals now have the law on their side, the future of farmed
fish and the wildlife around them remains uncertain.

STOLEN FISH

Bustling beaches play out a scene reminiscent of a coastal invasion.
A flotilla of long wooden boats, each crewed by half a dozen or
more men, makes for the breakers. White-painted bows rise and
fall with the incoming tide. Shouts of the crowds are muffled by
the deafening roar of the ocean. Women dressed in yellow, pink
and brown scurry into the shallows with large round washing-up
bowls, weaving between exhausted men to greet the catch, while
excited children dart around. Some of the women reconvene on the
beach to buy fish for smoking and to sell in the local markets. That
was once all there was, but now there is a new game in town: the
fishmeal factories, imposing white infrastructures just up from
the beach. A constant procession of fishermen heads to them from
the boats, carrying grey plastic crates. It's heavy, awkward work.
With expressions of resigned intent, they move up the beach to
feed the waiting belly of the factory.

In recent years, five fishmeal factories have appeared along the
coast of West Africa, three of which are in the Gambia, one of the
continent's smallest, poorest and most densely populated countries.
Most of what is produced in these factories is destined for export
to China and Europe. Fishmeal is finding a growing market as feed
for industrially reared animals, particularly fish, poultry and pigs.
With demand for fishmeal and fish oil increasing, particularly in
China, West Africa has become a hub that now produces 7 per cent
of the world's supply.[30] In the Gambia, fishmeal factories process a
national staple known locally as 'bonga fish' – a small pelagic fish
that is important for the country's food security, 90 per cent of
what is caught being consumed locally.[31]

Compassion in World Farming collaborated with the film-maker
Gosia Juszczak in the production of *Stolen Fish*, a documentary
that tells the stories of people caught up in the poverty trap caused

by the arrival of fishmeal factories on the Gambian coastline. The fishmeal company Golden Lead was the first to begin operations in Gunjur in 2016. Two more majority-Chinese-owned factories have since opened in the fishing villages of Sanyang and Kartong, along a thirty-kilometre stretch of the south-western coastline.[32]

Since the factories appeared, traditional fishermen have struggled to compete with better equipped trawlers that fish for the factories. They are all going after bonga and sardinella, fish that have always been abundant in Gambian waters. In the town of Bakau, close to the mouth of the River Gambia, one local fisherman tells his story by drawing pictures in the sand. With an athletic build, dreadlocks, vest top and red baseball cap, Abou Saine explains that commercial fishing vessels are hoovering up fish that congregate in the mouth of the river to breed during their homeward migration.

'This is the exact place where the Chinese fishing trawlers will come at night and steal the fish', Saine says. His face looks pained and his weary bloodshot eyes look incredulous as he talks about how the lights of fishing vessels gather at the mouth of the river at the same time each night. Small at first, then bigger as they get closer. They are in a part of the ocean Saine describes as being 'very much illegal for them to fish'. His account chimes with a study concluding that most fishmeal comes from fisheries in regions with low levels of governance, where pressure on fish levels can be high and the ecological impacts extreme.[33] In 2015, a Greenpeace exposé found seventy-four Chinese fishing vessels operating in prohibited waters off West Africa and falsifying vessel tonnages.[34]

'Before, you'd just stand here and see fish jumping, but now it's no more', Saine says. He believes that the fish have been taken away by 'overfishing with these big nets'. China's commercial vessels could take in a day what thirty local boats would take in a month. Before the fishmeal factories were built, Saine could catch enough fish to make a living in just a few hours; now, he has to fish for a week and still can't catch enough to live on. All of which has prompted him to consider a change in profession.

'I was thinking a lot that if I sell my boat, I could then at least try elsewhere, like Europe', he says. He would go the 'backway' – in other words, he would travel illegally. With a rapidly growing population, the Gambia has one of the highest rates of people travelling to Europe in this way.[35] His brother has been living in Madrid for the past fifteen years. The poverty here splits families. 'All I want is to see him again. Hug him, feel him, look at him, chat.'

Someone who knows the risks involved in 'taking the backway' is Paul John Kamony, a local fisherman who fled for a better life for himself and his family. 'As the first son, you are expected to take your family out of poverty. If I see a way to go to Europe, I will try it again', he says.

Kamony shaded his eyes from the sun and stood tall among his fellow crew-mates as the boat lilted to the rock of gentle waves. Wearing yellow waterproof over-trousers held up by thick braces, he hopped nimbly around the narrow vessel and helped to haul fine netting from the sea.

He would sell his catch to local women, who would then sell it on at the local market, and thus two vital links in the local economic chain sustained people who would otherwise teeter on the brink of poverty. About a dozen fishermen were in the boat, their arms rowing back and forth as they pitched in to haul the catch aboard. When it was in, the few fish they caught looked lost in the bottom of the boat. Disappointed, Kamony took a drag on his cigarette before starting to repair the holes in a tangled net. Soon, they'd motor off to try their luck elsewhere.

Most of the men he'd worked with over the years had left for Europe, and Kamony once tried the 'back way' himself. He asked his father for money to help begin a journey in search of a better life and got as far as Libya, where he worked in construction and tried to scrape together enough money to make it to Europe. He and a friend attempted to reach a boat heading to Italy, and that was when they were caught. Arrested and imprisoned for eighteen months, Kamony recalled seeing twenty-two prisoners squeezed into a single three-metre cell. Children, elderly people and even pregnant women, all caught trying to get to Europe. Unlike their fish bound for Europe in the form of fishmeal, they were not welcome.

'I saw many people die with my own eyes. Some were locked in a container until they suffocated', Kamony said. Four of his friends died when their dingy sunk on the way to Italy. 'I saw many people badly beaten. I was lucky, I was tortured only once', he said.

Many boats in the area now sell to the local fish factory, and one fisherman reaping the short-term benefits of the fishmeal boom is thirty-three-year-old Musa Duboe. Before the factories, he would catch red snapper and barracuda and sell it at the local market, but his income dwindled due to depleted stocks. 'Now work is booming again, as we can sell our catches to both the factory and locals', he told the *Guardian*. 'Our net catches all kinds of fish. Sometimes we meet demand with just one catch and other times we need five catches, with a catch being as big as up to 400 bowls of fish. My work is a lot more profitable and I can fully provide for my family, because the factory buys more fish than I could previously sell on the local market.'[36]

However, the reality is that selling fish for fishmeal both undercuts local markets and deprives hungry local people of fish, which provides around half the animal protein consumed in the Gambia.[37] Some proponents of fishmeal suggest that the small fish used, species such as anchovies, sardines, mackerel and herring, have little value as human food, but a recent study suggests that 90 per cent of fish used for fishmeal could be eaten.[38] Grinding these fish down to make feed for factory-farmed animals is therefore in direct contravention of the UN FAO's sustainable fisheries guidelines, which stipulate that fish should be used for feeding people directly. The code calls on countries to 'encourage the use of fish for human consumption',[39] and fishmeal clearly falls outside of this guidance.

In the Gambia, locals who once enjoyed fish as part of their daily diet are now struggling to afford it, while vast quantities of fish are instead being processed to feed farmed fish. It takes five kilograms of fresh fish to make one kilogram of fishmeal, with the resulting product being used as feed for farmed fish (69 per cent), pigs (23 per cent), chickens (5 per cent) and pets (3 per cent).[40]

As well as squandering hard-pressed fish stocks, fishmeal factories don't provide much local employment; one estimate suggests that the number of jobs created by each new plant is as low as thirty.[41] And with three plants in a country with a population of over 2 million, those jobs won't go far.

As fishermen rush to sell their catches to the new factories, reports have surfaced of overfishing to the extent that even the factories sometimes refuse to take them, causing unwanted fish to be dumped.[42] 'We are seeing dead fish thrown back into the sea, causing massive environmental pollution', said Sulayman Bojang, a small-business entrepreneur and local activist with the Gunjur Youth Movement.[43] 'We want to stop exploitation at the hands of the fishmeal plants, but with the Gambia being one of the poorest countries in the world, we stand no chance against the Chinese corporations.' The sum total of all this is a fishing industry under pressure, with the UN FAO calling for the once abundant sardinella fishery off the west coast of Africa to be reduced.[44]

Life for those who live from the sea has been made harder by the fishmeal factories, and this includes those who sell fish in the market. Mariama Jatta is a fish processor and market seller who lives in the port of Gunjur with eight children of her own, and two more that she takes care of. Wearing a floor-length blue-and-red dress, she stood commandingly on the busy beach, a pink headdress protecting her from the hot sun. Three local fishing boats swayed uneasily beyond the breakers. Other women, one with a small child on her back, clustered around large circular bowls. Grey-headed gulls clamoured to grab any fish that were spilt.

Jatta has worked in the port for forty years and admits to struggling to survive. She takes care of her large family alone so she carried fish, helped haul boats out of the water and scrubbed catfish, butterfish and bonga. She would buy fish for smoking and selling on the local market. The fishermen, she said, were exhausted. Their catch was meagre, and she was angry. 'Since the factory opened, the amount of fish we can buy for smoking reduced from ten to five baskets a day. All over the Gambia, fish prices went up', she said. 'The fishermen supply the factory first. Only when the factory's stomach is full, that's when we can get any fish.'

STARVING SEABIRDS

Waves of elation washed over me, as I stepped into a colony of a million birds for the first time. I can still feel it now, my senses overwhelmed by the intensity. I was soaking in one of most profound wildlife wonders of the natural world.

These were sooty terns returning to breed on Bird Island, a coral reef on the northern tip of the Seychelles archipelago, and they were nothing if not social. The concept of individuality blurred as I simultaneously tried to focus on a single bird and also to take in the entire spectacle. They were everywhere: in the air, covering the ground, right beside me and as far as I could see. It was an almost spiritual experience.

On Bird Island, sooty terns cover the flat grassy plain that's known as the 'Sooty Tern Reserve'. Smaller, more slender and swallow-like than gulls, with sharp beaks and deeply forked tails, they zip elegantly past your face and then settle back down to nest. Above all, it's the sound that I'll never forget. An electric pulse. A rush of noise hitting senses from all directions. Each individual bird has a nasal, almost electrified shriek with an upward inflection: 'way-awak!' Together, a million of them are literally deafening – noise levels at the tern colony can hit 107 decibels,[45] and employers are obliged to provide ear defenders to employees at 85 decibels. In the 1970s, Deep Purple became the world's 'loudest band' when they pumped out 117 decibels – not much higher than the terns – at a gig where three members of the audience were said to have fallen unconscious.[46] Standing with a million shrieking terns, I could feel why.

Soaking in the spectacle, I suddenly noticed intruders. Two people in khaki had clearly ignored the warning signs and 'KEEP OUT' notices. I could feel my blood rising. Then, when I thought things couldn't get any worse, one of them started lunging with a pole at grounded birds. I raised my binoculars and peered straight at them, a gesture I hoped they would take as me giving them a final warning. At the end of the pole there was a net containing a bird, all bunched and crumpled. With my patience now exhausted,

I set off to intercept the intruders. And that was how I first met scientists Chris Feare and Christine Larose. Feare had been coming to the island since 1971, when his post-doctoral fellowship brought him to undertake a study of sooty tern biology. Larose was his long-standing helper.

Half of the terns' eggs are harvested for human consumption and Feare was keen to discover whether that was sustainable. Taking some eggs and protecting the world's largest accessible sooty tern colony in return seemed a fair trade-off, and this is what Feare's studies have looked at. These terns are able to come and go at will, and the sacrifice of some of their eggs seems a small price to pay compared with the plight of hens on commercial farms. Whether they are free range or battery, commercial egg production means that male chicks of the laying strain are gassed or macerated when they are a day old. And when the hens are worn out, having been forced to pump out three hundred eggs a year, they are sent for slaughter. It's an end that happens for many at just eighteen months old, even though the normal lifespan of a hen can be six to eight years.

In centuries past, sailors would invade Bird Island to gather fresh eggs, deliberately trampling on all the eggs in the colony so they would know that any they found the next day were newly laid.[47] As well as worrying about egg-collecting, Feare was trying to learn more about where the sooty terns fed when out at sea. He was catching them on their nests in butterfly nets before adding a satellite tracker to the would-be parent.

Bird Island is one of the earliest examples of genuine ecotourism. Since the present owners bought the island in 1967, the terns have increased in number from 18,000 pairs to at least half a million pairs – a conservation success story. Although they are officially listed as a species of 'least concern' by the International Union for Conservation of Nature, the overall population trend for the sooty tern is described as 'unknown' and scientists have started to call for their status to be reviewed.[48] To Feare, such statistics hide the fact that many colonies have disappeared and been erased from the historical record.[49] He was putting satellite tags on the birds to see where in the ocean they foraged for food in order to work out how they were faring. They are long-lived, which means that it may be some years before anything affecting their breeding success becomes obvious. By the time a slump in breeding success affects the numbers attending the colony, it could be catastrophic.

Sooty terns face a multitude of threats, including accidentally being caught by fishing gear, the destruction of their breeding habitat and plastic pollution.[50] Climate change is already starting to affect them: variations in sea surface temperature make fish harder to find and lower the terns' breeding success.[51] Overfishing is also threatening their future; there have been significant reductions of tuna, a fish the terns rely on to push their prey to the surface.[52] Commercial fishing around the Seychelles is almost entirely for tuna (in 2016, out of a total catch of nearly 127,000 tonnes, nearly 121,000 tonnes was tuna).[53] Up to forty-six European Union fishing boats, mainly from Spain and France, fish for tuna around the Seychelles.[54]

Sooty terns breed on islands across most of the world's tropical oceans, living on a wide variety of small fish and squid.[55] During their stay on Bird Island, they make regular forays in search of fish to feed their young, journeys that were previously thought to range up to a hundred kilometres from their breeding grounds.[56] However, Feare's recent work with GPS trackers has found that adult birds can fly up to 250 kilometres from their nest. In extreme cases, he's found that a parent bird has covered more than 2,000 kilometres to find food, leaving their mate alone on the nest for up to thirteen days. Such evidence suggests that food is becoming harder to find.[57]

While the decline of predatory fish like tuna may partly explain why the terns are struggling, a general decline in the numbers of small fish must also surely play a part.

Seabirds the world over are finding it harder to find food, due to overfishing and the vast quantities of fish that are hauled from the oceans to be ground into fishmeal. Small fish are the mainstay food supply powering the world's ocean ecosystems, from predatory fish like tuna, to dolphins, whales and seabirds. Species like anchovies and sardines are not only important prey species for sooty terns, but also support much of the ecosystem. They feed on plankton and provide a bridge between the tiniest oceanic life forms and the larger creatures.

Small pelagic fish are a prime target of the fishing industry as well. Also known as 'forage' or 'trash' fish, they are used in animal feed, fertiliser and industrial oil. Anchovies and sardines are the mainstay in fishmeal, and recent figures show that these two species form more than half of the world's forage fish catch. They favour nutrient-rich waters such the Benguela upwelling off the coast of southern Africa or the Humboldt off the Pacific coast of Peru. Here, they breed in vast numbers, but have been caught in such quantities for so long that their numbers are depleting. And if small fish are no longer there, sooty terns and their chicks go hungry.

Sooty terns are in competition with fishing boats for fish. Of the 90 million tonnes of fish that are caught every year, 15 million tonnes are destined for fishmeal, and nearly three-quarters of the world's fishmeal is fed to farmed fish. China is by far the biggest importer of fishmeal, accounting for more than a million tonnes a year, followed by Norway, Japan and Germany, which each import about 200,000 tonnes a year. By comparison, the UK imports around 66,000 tonnes annually.[58]

Sooty terns are just one of many species of seabird that rely on small fish to survive, and seabird numbers worldwide have declined by 70 per cent since 1950. Commercial fishing has been a major cause of this decline, posing a threat that is described by scientists as 'substantial and global'.[59] On the UK overseas territory of Ascension Island, overfishing has led to sooty tern declines of more than 80 per cent.[60] Research suggests that industrial fishing and climate change have combined to rob them of their usual diet, leading them to resort to relatively low-nutrient food such as squid, snails and locusts.[61]

Colonies of related Arctic terns around Shetland and Orkney in Britain were particularly hard hit in the 1980s, when numbers of sand eels collapsed due to overfishing.[62] Terns are already declining because of competition from fisheries for food.[63] The bottom line is that all over the world, fisheries aimed at feeding factory farms are depleting the oceans and causing seabirds to starve.[64]

ANTARCTIC KRILL

Albert Einstein once said, 'Look deep into nature and you will understand everything better,' and nowhere does this apply more than to the evocative landscapes of Antarctica. In the last great wilderness on Earth, whales, seals and penguins grab the headlines. Blue whales, the largest creatures to have ever lived, are joined by humpback, fin and minke whales. Crabeater and Antarctic fur seals can be found by the million. Chinstrap, Adélie and gentoo are just three species of penguin that live and breed in vast numbers. Relatively few species thrive in huge numbers on an ecological cliff edge that is prone to boom and bust.

Antarctica's pristine wilderness has been tainted by merciless exploitation; in recent centuries, many of its seals and whales were hunted to the brink of extinction. Reports of vast numbers of whales and seals drew the adventurous to Antarctica from the early nineteenth century, and it wasn't long before humanity had overdone it. The Antarctic fur seal, for example, was almost wiped out. In more enlightened times, Antarctica's wildlife seems to be doing well. But just as populations of seals and whales are recovering, a new storm is gathering: krill fishing to feed factory farms.

The secret to the abundance of wildlife in Antarctica is krill, a small shrimp-like crustacean that grows up to six centimetres long. The environmental writer Kenneth Brower described them as appearing 'delicately carved from translucent, rose-tinted crystal'.[65] Among eighty-five species worldwide, *Euphausia superba* is the king of krill; in terms of biomass, they collectively weigh twice as much as humanity. They throng in 'swarms' of up to 10,000 per cubic metre, at such density that they turn the ocean pinkish-red. Tiny filter-feeders, krill are at the bottom of the food pyramid, feeding on abundant phytoplankton, thus making otherwise unobtainable nutrients available to the rest of the ecosystem. The mighty blue whale owes its existence to krill, each whale eating as much as four tons of the creatures a day.[66]

At first glance, krill seem super-abundant to the point of being indomitable. The Commission for the Conservation of Antarctic Marine Living Resources estimates that the biomass of Antarctic krill in the Southern Ocean is 379 million tonnes.[67] Krill fishing quotas amount to 3.7 million tonnes.[68] However, half the krill population is eaten by Antarctica's native wildlife. Crabeater seals and Antarctic fur seals depend on krill, as do gentoo, Adélie and emperor penguins. Albatrosses, petrels, prions, shearwaters and fulmars feed directly on krill or on things that eat them.

The balance is a fine one, but the twin forces of climate change and industrial fishing look set to upset it forever.

Scientists have already flagged a long-term decline in Southern Ocean krill. In a study led by Angus Atkinson of the British Antarctic Survey and published in *Nature*, eighty years of net sampling data was analysed, with the results showing that since the 1970s, krill numbers have suffered declines of 80 per cent in the regions of the Southern Ocean where they are most abundant.[69]

'Krill are dependent on the sea ice in winter', Martin Collins from the British Antarctic Survey told *National Geographic*. 'The juvenile krill in particular feed under the sea ice, and some of the adult krill as well. Declines in sea ice associated with warming means less habitat for krill and, therefore, in the longer term, certainly less krill … I think Atkinson's work suggests that decline is already beginning to happen.'

During the same period, krill fishing has increased more than six-fold, with Norway, China and Korea together accounting for more than 90 per cent of the total catch.[70] Antarctic krill can be dried, ground or used in oil form, with the latter marketed as a 'superior ingredient in aquaculture feed'.[71] It is also sold as a human health supplement and claims to be high in omega-3 fatty acids.[72]

The Norwegian company Aker BioMarine is the largest single krill fishing company.[73] In 2018, following a Greenpeace campaign, Aker was one of a number of companies that agreed to stop fishing in large areas of the Antarctic.[74] Although some may argue that the krill fishery catch is small, its impact on wildlife is already being felt. A recent study on the Antarctic Peninsula has shown that, as well as climate change, intensive fishing is making life harder for penguins.

George Watters, director of the National Oceanic and Atmospheric Administration, led a scientific analysis of more than thirty years of data on penguin performance, and found evidence that krill fishing has harmed penguins at about the same level as climate change. 'Our results ... are seeing impacts on penguins in the smaller areas where fishing actually concentrates', said Watters, suggesting that protective measures to prevent overfishing were not going far enough.[75]

Commenting on the study, Professor Alex Rogers, a specialist in sustainable oceans at Oxford University, said: 'Climate change is having an impact [on the krill population] at the same time as a resurgence in krill fishing, with increasing numbers of fishing vessels and changes in technology that are causing increasing krill catches.'[76]

Conservationists have been warning of declines for some time. The International Union for Conservation of Nature has warned that large-scale krill fishing poses a threat to Antarctica's crabeater seals, one of the world's most numerous remaining large mammals.[77] Antarctic fur seals, which feed primarily on krill, could also be affected by the expansion of fisheries,[78] as could Adélie and chinstrap penguins. The *Handbook of the Birds of the World* predicts that the overfishing of krill 'could have a catastrophic effect on the food chains of the Southern Ocean'.[79]

For now, at least, agreements are in place to lessen the impact. A global catch quota amounts to 1 per cent of the estimated volume of krill, while voluntary restrictions aim to limit fishing around penguin breeding colonies. Yet the writing must surely be on the wall: as commercial fish stocks fail around the world, the temptation to scale up Antarctic krill fishing is likely to become irresistible.

According to the UN, 90 per cent of the world's marine fish stocks are now fully exploited, overexploited or depleted, and subsidies of more than $20 billion a year that favour big industrial fleets over small-scale fisheries have contributed to the problem.[80] At the same time, the FAO has suggested that the inevitable gap between demand and supply could be plugged by fish farming,[81] an assessment that overlooks the fact that the farming of carnivorous species such as salmon, trout and octopus, is, in fact, a net consumer of fish. Far from solving the world's fishery problems, fish farming can simply add to them.

As fisheries dry up, krill fishing for fishmeal is likely to escalate, and so the long tentacles of factory farming extend yet further around the world, invading its last wilderness. While overfishing is often met with a shrug – after all, we've got to feed people – few make the link between fisheries and industrial agriculture. In Antarctica, the pace of decline is quickening. If we are to preserve life on Earth, we should take Einstein's words to heart and 'look deep into nature' while we still can.

PART THREE

WINTER

Winters in my part of southern England are more wet than white. A covering of snow is rare and usually fleeting; hard frosts are the closest we get to that Christmas-card scene. Today there was relentless rain and gusting wind. When the wind dropped, we headed out, with our dog Duke bounding ecstatically through sprawling puddles. We walked past the badger sett. Not far from here, I'd previously seen a rare all-white badger disappearing into the undergrowth, an extraordinary sight. We often see the family of black-and-white-striped ones playing on the slopes, but today they stayed out of sight. Some of the entrance holes to their sett had been blocked, which is illegal. The landowner was dismissive, just left them blocked and didn't want to report it, so I did. As yet, there's been no repeat.

Despite winter biting with a vengeance, Britain's smallest bird was brave enough to venture into song, his thin whistles bringing to mind a delicate lasso swishing in the air. There, hopping from branch to branch, was an exquisite goldcrest, green and stripey with a yellowy crown. A flock of hardy redwings moved in waves as we approached. Beyond that, things were quiet; wildlife had hunkered down.

With the rain lashing down, soil was on the move, brown water running over exposed roots and gushing down slopes. Where it washed off roadside fields of maize stubble, it was heavy with mud which swept into the carriageway.

As for the river, the Rother was riding high and broken branches were swept along like twigs. Bankside willows that were usually dangling well above the river now dipped into the surface and reminded me of an outstretched hand, its fingers touching the water. A footbridge to an alder copse was submerged, with only the handrail visible. If the river were a foot higher, the banks would overflow; I looked across to our hamlet with worry. The day was darkened by a thick, suffocating blanket of cloud. Rain continued to fall, and the river continued to rise, while the cows bellowed in their winter barn.

Then the river burst its banks and the wash ran across the fields like an incoming tide. We watched it heading towards our home; much higher and we would need sandbags. The arches of the stone road bridge beside our hamlet were submerged. Dark flows of muddy water twisted in search of somewhere to go, and the grazing meadow flooded. When the wind dropped, things looked strangely beautiful, but as dusk fell, the rain swept back in. That night, we didn't get much sleep. It was an anxious time, not knowing what we might wake up to…

10

THE DUST BOWL ERA

PROSPERITY THIS WAY

'Riches in the soil, prosperity in the air, progress everywhere',[1] was the promise of syndicates and speculators, pamphlets and posters. People moved to the American Great Plains in search of affluence, clean air and a fresh start. Hundreds of thousands settled on the endless, flat grasslands, prompted by the biggest government giveaway of land in American history[2] – 160 acres per homestead, yours forever if you could hold down the land for five years. They came from Europe and the east coast of America, driven by the promise of a better life. Among them were farmers, former slaves and single women.[3] For them, making a fortune was not the only thing on offer; there was also freedom from discrimination.

One of the single women was Caroline Henderson, a schoolteacher and farmer's daughter from Iowa who arrived on the plains of Oklahoma in 1908. She settled thirty miles east of Boise City in a little place called Eva. A year earlier, she had been struck down by a near-fatal dose of diphtheria. Having recovered, she decided to follow her dream of a future on a western ranch in the American West. Her calling was inspired by the 'Jeffersonian dream', an idea espoused by the US Founding Father Thomas Jefferson that the country should be a nation of agrarian smallholders owning their own land. According to Jefferson a century earlier, the route to finding land to fulfil this vision was westward expansion.[4]

Soon after staking her claim on the plains, Caroline wrote enthusiastically to a lifelong friend: 'So here I am, away out in that narrow strip of Oklahoma between Kansas and the Panhandle of Texas, "holding down" one of the prettiest claims in the Beaver County strip. I wish you could see this wide, free western country, with its great stretches of almost level prairie, covered with the thick, short buffalo grass, the marvelous glory of its sunrises and sunsets, the brilliancy of its starlit sky at night.'[5]

Born in 1877 to a wealthy farming family in Iowa, Caroline become one of the most remarkable writers of frontier history, describing her westward adventure and the daily hardships facing those living through the most difficult of times, first in letters to friends and family and then in articles published by *Ladies' World*, *Atlantic Monthly* and *Practical Farmer*.

Caroline arrived in Oklahoma having just turned thirty, and within months started teaching at a local school. Here, she worked while building her first home – little more than a cabin – on the homestead where she would live for six decades. Among the crew she hired to dig a well to help satisfy the government of her claim was a tall, dark, moustached man called Wilhelmine Eugine Henderson. His grey-blue eyes stood out from skin browned from years living on the plains. He was gentle, optimistic and straightforward. They were married in May 1908. She remembered it as 'one of the most perfect days'. Will remembered the well as the best he'd ever dug. So, he decided to keep it, and the schoolteacher who went with it! Their first and only child, Sarah Eleanor, was born in 1910. They shared a love of the land, rearing turkeys, chickens, cattle and pigs, ploughing the land and planting wheat. They were enchanted by the land, tested by it but never broken.

The southern Great Plains where Caroline settled were seen as agriculture's last frontier; a flat, treeless expanse of buffalo grass with constant wind and low, unpredictable rainfall. Impetus for pioneering settlement was driven by manifest destiny, the belief in spreading democracy and capitalism westward across the country.[6] Many settlers lived by the superstition that 'rain follows the plow', and they were not alone. Politicians, land speculators and even some scientists believed that setting up homesteads and breaking the land would change the otherwise arid climate, bringing better conditions for farming. These were 'next year' people willing to farm in the face of adversity, their future defined by the words 'if it rains'.[7]

Hope, greed and then poverty drove settlement on the Great Plains, but this was preceded by a struggle between the white pioneers drawn westward and the Native Americans who had long roamed this vast, unforgiving landscape in pursuit of bison. The seeds of one of the world's great man-made environmental disasters were sown when millions of bison were replaced by millions of cattle and then millions of acres of ploughed cropland. The contract between settlers and the soil had been torn up in spectacular style.

For millions of years, the roots of the perennial grasses of the Great Plains had held together an otherwise fragile landscape. In Texas alone, over 470 native species of grassland carpeted two-thirds of the state.[8] A natural ecosystem was kept in balance, the vulnerable soils bound by a sprawling mass of roots and cloaked by a thick mat of shoots. Plains grasses may look scraggy and dead during dry times, but beneath the surface there is a deep root system that can survive anything but the most prolonged drought. In the face of adversity, it is what the environmental historian Donald Worster called one of 'nature's winning designs.'[9] This symphony of photosynthesis sustained all manner of wildlife, not least the most incredible herds of bison.

WILD GRAZERS

At their peak, between 30 and 50 million bison roamed the Great Plains of America, collectively weighing about the same as the entire human population of North America today. Sustained by nothing more than sunshine, rain and grass, these vast herds exemplified grazing animals living harmoniously with their environment.

After the American Civil War, European settlers began to move west. New army posts and railroads sprung up, and hunters were contracted for the lucrative business of supplying buffalo meat for soldiers and workers. Hunters arrived with high-powered rifles and each killed as many as 250 animals in a day. The slaughter reached its peak in the 1870s, when 7 million pounds of bison tongues were sold in two years.[10] As the killing got out of hand, sportsmen spread across the plains to kill bison for fun. Railroads began advertising 'hunting by rail'; people would pay to shoot herds indiscriminately, with the downed animals simply left to rot.[11] According to one estimate, some 25 million bison were slaughtered in 1872 and 1873.

For centuries, bison had sustained the Indian tribes who relied on them for food, clothing, tools and shelter. Within a decade, Indian cultures were confined to reservations while bison were facing extinction. Settlers had won the day; the rich grasslands were conquered for cattle and a wildfire of change swept the plains. Yet the climate on the Great Plains has always been unforgiving; it is largely arid, with less than twenty inches of rainfall in an average year.[12] Indeed, the plains east of the Rocky Mountains became known as the 'Great American Desert'.[13]

Following a few unusually wet years early in the twentieth century, settlers threw caution to the wind. Encouraged by high grain prices during the First World War, they ploughed millions of acres of grass cover and replaced them with cereals, mainly wheat. In the 1920s, millions of acres of grassland disappeared in what became known as 'the great plow-up'. Tractors were introduced, as settlers produced record amounts of crops and animals. However, the subsequent surplus led to a crash in prices.

Settlers were joined by 'suitcase farmers' trying to 'hit a crop'. They had no intention of staying but came to plough up grassland, plant their crop and come back months later to bank the harvest.[14] Caroline Henderson described them as having 'bet thousands of dollars upon *rain* ... [in] the preparation of large areas of land all around us which no longer represent the idea of *homes* at all, but just parts of a potential factory for the low-cost production of wheat – *if it rains*'.[15]

Soil was seen as inexhaustible, an attitude normalised by the Federal Bureau of Soils, which called it 'the one indestructible, immutable asset that the nation possesses. It is the one resource that cannot be exhausted, that cannot be used up.'

During the 1920s, wheat production soared by 300 per cent,[16] and both crops and animal products saw record yields. But supply exceeded demand and prices slumped, after which the Wall Street Crash of 1929 dragged America into the Great Depression. Crop growing turned to factory farming, with agricultural holdings increasingly seen as food 'factories'. Wheat was particularly suited to farming's new industrial approach. And it was scalable; the more land you could dedicate to it, the more money you'd make. This was the 'Henry Ford model' brought to agriculture,[17] the application of industrial capital to a farm setting.

By 1931, the wheat harvest had broken all records and its production seemed limitless.[18] Caroline Henderson and her husband had also been swept along by the wheat bonanza. That summer, they too managed a bumper harvest. The national wheat harvest was so huge it caused a glut and grain stores overflowed. When prices crashed, farmers struggled to recover the money it had cost them to produce the crop. 'We are too big to cry about it, and it hurts us too much to laugh,' Henderson wrote. They had at least bought a tractor, a combine and a new house without getting into debt, but, as Henderson noted, others weren't so fortunate. 'What of the people who are paying rent, who are in debt for legitimate expenses, who have the expense of illness or larger families to provide for and educate?'[19]

Homestead settlers found themselves having to decide whether to cut production to boost prices or to produce more to make up for falling returns; desperation drove them to choose the latter. An army of tractors stripped the grass from fragile soils at the rate of 50,000 acres a day.[20] They banked on the plough having changed the Great Plains climate forever and didn't foresee eight long years of drought. For now, they kept on ploughing.

The rich covering of perennial grasses that once protected the foundation of this fragile ecosystem – the soil – were upended to make way for the crop. The stage was set for environmental and economic disaster.

Drought set in. During the autumn of 1931, a lack of rain meant newly sown crops failed to grow, leaving soil exposed.[21] Fields that had been green dried up. Despite the drought, ploughing continued regardless, as people clung to the mythical belief that 'rain follows the plow'. By cutting the ground, they thought they would change the climate for the better.[22] It was a huge gamble, but they had little choice. They had debts to pay on new tractors, combines, ploughs and land as well as taxes and the cost of living – producing more was the only way to make up for crashing prices. By the early 1930s prices were so low that many of them went bankrupt.[23] And when prices fell still further, the 'suitcase farmers' simply abandoned their fields.[24]

Millions of acres of former grassland were now laid bare, exposed to winds that blew fiercely across a dry and dusty land, causing large chunks to crack and break. 'It's a very complex ecosystem that evolved over thousands of years,' wrote Timothy Egan. 'In less than a generation's time, they peeled the whole thing off.'

DUSTERS

At around midday on a windswept day in January 1932, a dust cloud rose outside Amarillo, Texas. There had been dust clouds before but none as big as this; 10,000 feet high, it rolled across the city, blocking out the light. Locals struggled to describe it. It wasn't a rain cloud or hail; it was thick, like coarse animal hair, a blizzard with an edge like steel wool. When it hit, day turned to night; you couldn't see your hand in front of your face. Streets were strewn with black dust so pervasive it got inside homes. The dust got into people's hair and throats, burned their eyes and made them cough.[25]

Amarillo had been hit by the first of the 'dusters', and from then until the end of the decade, these devastating storms became a frequent event. Shaped like enormous cumulus clouds, they would hug the ground, rolling on themselves from the crest downward. More than a dozen were recorded in 1932, with many more the following year. By 1934, it was estimated that 100 million acres of farmland had lost all or most of its topsoil to the winds.[26] Western

Kansas, eastern Colorado, north-eastern New Mexico and the Oklahoma and Texas panhandles were hardest hit, with the region becoming known as the Dust Bowl.[27]

'I didn't know what was coming,' recalled Lorene Delay White, a child when one of those early storms hit her home in south-western Kansas. 'It got darker and I was really afraid. Before it hit us, rabbits ran ahead of it, birds flew ahead of it. And we had never seen anything like it … And then it finally rolled in on us. You couldn't see anything.'[28]

For Caroline Henderson, the relentless nature of the dust storms took her to breaking point. On some days, the dust would be so bad she couldn't even see the windows in her home because of the 'solid blackness' of the raging storm. The dust would get everywhere – she described having 'Dust to eat, and dust to breathe and dust to drink.' Hopes of a decent harvest were snuffed out by the dust, leaving her facing a fourth year of crop failure. Grazing animals were no longer an option, as native grasses had been smothered by drifting dust.[29]

The Dust Bowl turned out to be a decade-long apocalypse. Without the water-retaining roots of native grasses, exposed topsoil blew away. 'Black blizzards' of windblown soil blocked the sun, filled the sky and marched across the landscape like tornadoes on their side. Dust piled high in drifts. By the mid-thirties, scientists would estimate that the southern plains alone had lost 850 million tonnes of topsoil to the winds.[30]

Stories abound of the people who were killed by the storms. A farmer whose car left the road in a black blizzard got out to walk the two miles home, but never made it. A seven-year-old boy in Kansas got caught in a storm and was later found dead in the dirt.[31] Two cars smashed head-on while crawling through blinding dust, killing both drivers.[32]

As things went from bad to worse, many dug deep and stuck it out whereas others headed for a new life elsewhere in America. Some 2.5 million people left the Dust Bowl states during the 1930s, the largest mass migration in American history. Some 440,000 poverty-stricken people left Oklahoma in search of better times, a story immortalised in John Steinbeck's classic novel, *The Grapes of Wrath*.[33]

Dust storms also of course affected animals. Chickens in their hen houses were smothered by dust, while those caught out in the open were blown away.[34] In one Texas county alone, dust killed 90 per cent of the chickens. Cows stopped milking and cattle left out on the range went blind, their eyes cemented closed.[35] Others starved or died from 'dust fever', their stomachs lined with dirt.

Following the storms, hungry jackrabbits would swarm. 'They ate everything green there was', Dust Bowl survivor Dorothy Sturdivan Kleffman said. 'The farmers had killed off the coyotes and that upset the natural order of things, and the rabbits just exploded and they would eat anything green they found. They would eat your garden up.' Huge numbers of rabbits competed with people and cattle for whatever the dust hadn't destroyed. Farming communities organised 'rabbit drives', where the desperate animals were driven into a large pen before being beaten to death with clubs and bats. 'They would scream', remembered Dale Coen. 'I can still hear rabbits, the noise that they'd make. I went on one rabbit drive and that was enough for me.'[36]

However, not everyone saw jackrabbits as the enemy; Caroline Henderson took pity on these fellow victims of the environmental tragedy. One Easter morning, she found one sheltering in her kindling pile beside the kitchen door, trembling, starving and missing an eye. The rabbit made no attempt to escape, so she picked him up, bathed his eye and put him in with her pet guinea pig to recover. She wrote about her feeling of kinship with her fellow animals in a column for *Atlantic Monthly*: 'When these wild creatures, ordinarily so well able to take care of themselves, come seeking protection, their necessity indicates a cruel crisis for man and beast.'[37]

In the early 1930s, devastating storms on the plains were not the only thing America was struggling with; the nation had sunk deeper into economic depression. Crop prices had plummeted on the plains, but drought meant that there wasn't much to harvest anyway. In Kansas, Caroline Henderson described how, though little was growing, people had no choice but to carry on: 'In nothing that we can produce here is there at present the slightest chance of any return on our labour. Yet we keep on working – really harder than ever.'[38]

NEW HOPE

In March 1933, America swore in President Franklin D. Roosevelt, who addressed what he described as the dark realities of American capitalism[39] and the blackening storms on the prairies. He had grown up surrounded by nature in the Hudson River Valley, and it had left a conservationist streak within him.[40] The storms and drought left him deeply moved; he would later say, 'I shall never forget the fields of wheat so blasted by heat that they cannot be harvested. I shall never forget field after field of corn stunted, earless and stripped of leaves, for what the sun left the grasshoppers took. I saw brown pastures which would not keep a cow on fifty acres.'[41]

Roosevelt acted quickly to prevent the nation's food providers going bust, introducing government measures to tackle surpluses and stabilise farmers' incomes. The first farm bill, known as the Agriculture Adjustment Act, was passed by Congress in 1933 as part of his New Deal. In a bid to restore prices, farmers were paid to *not* grow food on part of their land, with surpluses bought by the government. State support for agriculture was introduced, something that persists to this day. Among those signing up was Caroline Henderson, who agreed to reduce her wheat planting by thirty-nine acres. 'We feel as if the administration is really making a sincere effort', she wrote.[42] Meat prices were also stabilised through a new federal programme to reduce the nationwide surplus of cattle and pigs.[43]

CONSERVATION

To stop the soils blowing away, Roosevelt summoned the conservationist Hugh Bennett to the White House. Forceful and charismatic, Bennett, who would become known as 'the father of soil conservation', set to work as director of the newly established Soil Erosion Service. His first job was to unpick the myth that soil was 'indestructible'.[44]

On 14 April 1935, the worst 'duster' of the Dust Bowl drought – 'Black Sunday' – marked the turning point in the plains' soil crisis. Until then, the storms had been distant and abstract, but this time the dust spread as far as the capital, where fine powdery sand left congressmen choking. Bennett was testifying before a congressional committee in Washington DC, and aware that the storm was coming, he used it to demonstrate the need for soil conservation. The resulting Soil Conservation Act of 27 April 1935 created the Soil Conservation Service at the United States Department of Agriculture.[45]

Bennett could see that the Dust Bowl catastrophe was man-made – he blamed 'mistaken public policies'[46] and a 'system of agriculture that could not be both permanent and prosperous'.[47] Convinced that conservation techniques could help rehabilitate the plains, he foresaw farmers rotating crops, fallowing land, ploughing in contours or wavy lines to stop rain washing away the land or abandoning the plough altogether.

From then on, soil conservation became a priority. Forty thousand farmers signed up to a package of incentives in return for adopting practices that would conserve soil, with the amount of erodible soil consequently being reduced by half. Grazing was regulated to stop the erosion of remaining grasslands, while millions of acres of farmland were bought by the government and returned to pasture. The Great Plains Shelterbelt Project saw huge resources put into planting over 200 million trees in the Dust Bowl states, a huge windbreak that would stem the winds and their destructive power; even today it is regarded as one of the greatest environmental success stories of all time.[48]

By 1939, the drought was all but over, thanks to Roosevelt's intervention and the return of rain. In a 1940 speech in the Great Smoky Mountains National Park, Roosevelt reflected, 'We slashed our forests, we used our soils, we encouraged floods ... all of this so greatly that we were brought rather suddenly to face the fact that unless we gave thought to the lives of our children and grandchildren, they would no longer be able to live and to improve upon our American way of life.'[49]

The Hendersons were once again reaping good harvests. 'Wheat was a fair crop … We had ample pasturage with the increased rainfall and cattle have done reasonably well … and we had a nice garden with most of our winter's living stored away.'[50] However, years of 'dust devils' had taken their toll, and Caroline had aged beyond her years. A local newspaper printed a photograph of her standing with her sister Susie but mistook them for mother and daughter. Alongside her youthful-looking sibling, Caroline was thin, wrinkled and worn.[51] She and Will continued farming their homestead well into the 1960s. They were finally forced to leave due to ill health, shortly before passing away within months of each other in 1966. What remains of their homestead has been placed in trust, meaning it will never again face the plough.

Roosevelt's approach to soil conservation involved facing facts and dealing with them, giving thought to the needs of generations to come. With environmental threats on the rise once again, the question remains: will those who now fill his shoes follow his example?

11

CELEBRITY FOOD VILLAINS

HOW BIG AG KILLED THE COUNTRYSIDE

It took me twenty-five years to pluck up the courage to say something that I had always thought was obvious: that the most efficient way to keep a cow is grazing on a grassy hillside. Yet for much of my career, it has been drowned out by a school of thought suggesting that cows should be fed on grain grown on prime cropland. Yet a cow on a grassy hillside would be grazing land that wasn't useful for crop production, or perhaps on land left to recuperate after crop production. And in eating grass, cows and other ruminant animals thrive on something that can't be eaten by people. In this way, farmed animals utilise marginal or rotational farmland while adding to the food basket. And to me, that's efficient.

However, saying this in the company of the modern farming industry has long risked ridicule. I remember receiving criticism on Twitter for daring to suggest that cows belong in fields, with farmers who used intensive methods pouring scorn on such a thought. Didn't I realise that cows would be much less productive if they were kept naturally? I got trolled mercilessly, to the point where my family were also targeted. I came home on several occasions to find my wife in tears, wounded by a torrent of abuse. A local radio station picked up the story, which unfortunately became about the dark side of social media; the plight of the countryside got lost.

I learned in that moment how difficult it can be to speak out about simple truths, especially when vested interests are threatened. I was fortunate to discover the Pasture-Fed Livestock Association (PFLA), a group of farmers who realise that cows belong in fields and who pledge to feed their animals on grass and other greenery. When it was established, it was a revelation in farming circles. Grazing animals on grass instead of grain – who would have thought? Since then, the organisation has expanded to include hundreds of farmers, and its message is resonating internationally among forward-thinking custodians of the countryside. I've found similar movements in the United States, Italy and Argentina, often encapsulated by the words 'pasture-fed', 'regenerative' or 'agroecology'. Whatever label you put on it, it's about treating animals like living creatures in a countryside that we depend on for our well-being.

Seeking the truth about food production and how we treat animals in the countryside has been the motivation behind my thirty-year-long journey to get under the skin of the industrial farming industry. My quest gained fresh impetus a decade ago. Travelling with the journalist Isabel Oakeshott, I set out to join the dots between animal cruelty and the impact on the environment and human health. Ten years on from my first book, *Farmageddon*, there is much greater awareness, as demonstrated by inspiring movements such as Extinction Rebellion and Greta Thunberg's school strikes for the climate. At the heart of the climate, nature and health emergencies is a reckless industrial approach to farming.

Achieving change is doubly hard when powerful vested interests stand behind the status quo. I decided to take a journey into the belly of the beast that is factory farming, to examine those vested interests that keep a deeply damaging system locked in.

INTO THE BELLY OF THE BEAST

There might have been better days to drive to the West Country of England, but where I was going, the weather made no difference – the conditions are always the same inside a battery hen shed. After a long drive, I turned onto a concrete track in a place where countryside met an industrial estate, with offices and large warehouses surrounded by fields. A coffee was waiting for me at reception and my hosts were welcoming – they wanted to make a good impression. I could see whatever I wanted, but they asked that I refrained from taking photos – they didn't want their business made public. I was about to see first-hand how most eggs are produced.

Walking into a windowless warehouse, I was overwhelmed by the sight of row after row of cages stacked on top of each other. The noise was deafening. Through the gloom, you could make out the frantic heads of hens flicking in and out of cage fronts, the only sign that the cages were inhabited by living creatures. The birds were fixated on granules of feed delivered by conveyor belt, the single point of interest for much of their day. They were crammed in so tightly that they couldn't stretch their wings. There were 100,000 hens locked within a mass of wire mesh, each bird with less space than a piece of standard A4 typing paper. They stood on bare wire, crouching slightly because the wire ceiling made standing upright impossible.

My invitation to see battery hen cages in Britain came in the 1990s, after I'd led an undercover exposé of battery eggs being sold in supermarkets as 'fresh', 'farm fresh' or 'country fresh', reassuring labels that belied the truth. The industry was keen to engage with me in a bid to head off further trouble.

Thankfully, the outcry that led to my visit, followed by decades of campaigning, meant that these 'barren' cages were banned in Europe in 2012, yet the so-called 'enriched' cages that replaced them aren't much better. Though they give hens a perch and a postcard-sized bit of extra space, a cage is still a cage. We've also seen progress on labelling: in Britain and Europe, eggs from hens reared in 'enriched' cages have to be labelled as 'eggs from caged hens', although the writing is usually as small as the industry can get away with.

Despite some reform, the industrial model of egg production prevails. Most eggs produced in Europe come from caged hens, while the proportion of caged hens in Britain has dropped below half, thanks to the work of Compassion in World Farming and others. An outright ban is long overdue.

When I joined Compassion in World Farming in 1990, I made it my business to see for myself the reality of how animals are kept, in order that I could talk convincingly about what was going on. I went to factory farms with an undercover investigative team, and I followed live animal export consignments from the Midlands to Dover and across the Channel to the Continent. I've never lost the drive to see what's going on and to find the hidden story behind those celebrities of the food world: chicken, eggs and bacon.

These are food's celebrity villains, akin to the Kray Twins, the infamous British criminals of the 1960s. Like Ronnie and Reggie whose glamorous lifestyle hid a deeply dark side, so the two classic ingredients in a 'Full English' – bacon and eggs – have backstories that marketeers go to great lengths to hide; and most of it is manufactured using the cruellest of systems. For meat especially, honest labels are hard to come by and the industry uses marketing spin in a desperate attempt to defend the indefensible. Assurance schemes, like the Red Tractor logo or the British Lion Mark on eggs in the UK, in fact assure little more than compliance with minimum legal requirements on animal welfare.

In recent years, I renewed my impetus to witness the realities of factory farming around the world and persuaded Isabel to come with me. It was 2011 and she was a new mother, so how to do best for her baby was uppermost in her mind. I wanted to show her what was really going on behind the opaque veil of food labels like 'fresh', 'farm fresh' and 'country fresh'.

We started in California, the spiritual home of the kind of mega-farming touted at the time as the way forward for farming in Britain and Europe. To Isabel's chagrin, we skipped the bright lights of Hollywood for Central Valley, where farming was on a scale and intensity I'd never seen before. There were massive fields of crops and huge plantations of almonds and fruit. The crops seemed to be sprayed incessantly with pesticides dispensed from helicopters or hi-tech vehicles driven by people wearing masks. The wild bees had disappeared. So 40 billion bees would be brought into the state every year by thousands of trucks to do what nature was no longer allowed to do: pollinate the crops. When we took to the air in a small plane to get a bird's-eye view, the sheer scale of this vast patchwork quilt was eye-popping. There were huge continuous fields of single crops, without hedgerows and unpunctuated by woodland. Every now and then, we would see what looked like a vicious scar on the landscape. These were mega-dairies, each with up to 12,000 cows in a single dusty paddock, and without a blade of grass in sight. When I told Isabel that this was what some people saw as a vision for the future of farming, she turned to me aghast and said, 'This looks more like farmageddon.'

If we draw back the veil on this scenario, we find that it's not farmers who are making money from it, but 'Big Ag'. Behind the marketing gloss and talk of 'efficiency', the businesses that make and sell animal feed, the chemicals needed to grow it, the equipment for keeping animals confined and the veterinary drugs to stave off inevitable disease are big money-spinners. Large multinational companies have sprung up to supply farmers with the trappings of 'modern' agriculture.

At its heart, factory farming relies on a reliable supply of cheap, plentiful grain. Crops destined to feed intensively reared animals tend to be grown intensively, which makes for a big market for pesticides and artificial fertilisers.

The rise of animal farming as a major market for grain harvests has its origins nearly a century ago, in the American Dust Bowl and the Wall Street Crash of 1929. When prices for agricultural goods plummeted, farmers boosted production, which caused prices to collapse further. Grain became so cheap and plentiful that it may as well be used as animal feed. The Farm Bill, a mammoth package of US government subsidies aimed at helping farmers struggling in the face of the Great Depression, only made the situation worse.

In modern times, the US Farm Bill describes its main subsidies as coverage for 'agricultural risk' and 'price loss', yet their effect is to make farmers less responsive to the needs of both market and climate. As a report by the Washington-based Environmental Working Group put it, the Farm Bill 'rewards farmers for not adapting to the changing climate. Instead, it encourages them to plant the same crops in the same way, year after year, pretending poor harvests don't happen and repeating the mistakes of the 1930s.'[2]

In Europe, the Common Agricultural Policy (CAP) introduced subsidies with the aim of protecting European farmers from international competition and price fluctuation,[3] but it distorted the market. Farmers produced more than was needed, resulting in the notorious butter and beef mountains and wine lakes of the 1980s. CAP payments incentivised large-scale arable production, leading to the rise of 'barley barons' – cereal producers who became rich on subsidies.

The CAP has long been criticised as leading to a shift towards intensification. The European Commission itself admits that it encouraged farmers to 'use modern machinery and new techniques, including chemical fertilisers and plant protection products',[4] the latter being a euphemism for pesticides. In more recent years, subsidies have been largely decoupled from production and instead based on the area farmed, but this has incentivised farms to get bigger still.

During the last decade, the EU was spending some €50 billion on CAP farm payments every year, accounting for about 40 per cent of the EU's entire budget.[5] Britain introduced its own subsidy scheme with the 1947 Agriculture Act, but this was replaced by the CAP when it joined the European Common Market in 1973. Before Brexit, British farmers received more than £3 billion in subsidies a year.

Alongside a boom in subsidies to agriculture, there has been a drop in the prices received by farmers coupled with a rise in the price of inputs – fertilisers, pesticides, machinery and so on. Much of the subsidy, then, goes to Big Ag. Just as happened in the US Dust Bowl era, farms get stuck on a treadmill, having to pedal faster to produce more at lower cost. Farms specialise in producing a limited number of goods, as scale becomes everything and intensive, chemical-based farming comes to dominate.

Cereal farmers need a globally expanding customer base if they are to increase sales, and that's where industrial animal farming comes in. If you grow cereals to feed people, your sales will be limited by population size. But if you sell wheat or corn to feed animals, you'll be able to sell more of your product to the same number of people. It takes several times the amount of wheat, corn or soya to provide an equivalent in meat; at best, for every hundred calories of wheat, corn or soya we feed to farmed animals, we receive just thirty calories back in the form of meat and dairy products. When it comes to protein, the loss in conversion is similarly stark. For every hundred grams of grain protein fed to animals, we get back about thirty-five in eggs, forty in chicken, ten in pork or five in beef.[6]

Squandering grain as animal feed is big business. The revenue for eleven leading international companies producing meat and feed grain is in the order of $250 billion per year. One of the largest is the CP Food Group in Thailand, which has sales revenues of around $18 billion from animal feed, livestock production and processed foods.[7] The largest Western company is Cargill, whose global operation spans animal feed, food ingredients and livestock production, and has total revenues of $113.5 billion.[8] Tyson Foods of Arkansas is a major producer of chicken, beef and pork, with 2019 sales revenues of $42.4 billion.[9] The enormous Brazilian meat company JBS operates all around the world, and can slaughter 14 million chickens per day.[10] It has a net operating revenue equivalent to $40 billion.[11]

As chief executive of a global charity fighting to end factory farming with campaign budgets that are comparatively minuscule, I'm under no illusion of the scale of the task in pressing for reform. And our task is made even harder by phoney rhetoric that factory farming is somehow 'efficient', when the reality is that feeding animals on grain wastes the vast majority of the crop's food value. By contrast, grazing cows and sheep on grassy hillsides or rotational pastures converts what we can't eat – grass – into something we can. Chickens and pigs foraging woodland edges or being fed properly treated food waste do the same.

For Big Ag, the 'efficiency' is in being able to sell feed, chemicals and drugs to farmers. The annual combined artificial fertiliser sales of just three companies – Nutrien,[12] Yara International[13] and the Mosaic Company[14] – are $42 billion.[15] Corteva Agriscience had worldwide sales of $14.3 billion in 2019.[16] The 'Crop Science' division of Bayer (which acquired the agrichemical company Monsanto, producer of GM herbicide-resistant crops and the herbicide Roundup) had annual sales of €14.3 billion.[17] Syngenta, one of the large number of companies owned by ChemChina, has sales of $13.6 billion.[18] Completing the set, BASF has agrichemical revenue of close to €7.8 billion.[19]

Back on the factory farm, the presence of so many animals in close proximity creates a melting pot for disease, providing Big Ag with the opportunity to sell antibiotics and other pharmaceuticals to farmers. It is no coincidence that 73 per cent of the world's stock of antibiotics goes to farmed animals, largely to ward off diseases associated with intensive farming.[20] The biggest six drug companies rely on the animal farming sector for $9 billion of sales each year.[21] These enormous numbers show how Big Ag thrives while animals and the countryside suffer, and ordinary farmers go to the wall.

PITY THE POOR GRAIN MERCHANT

If it weren't for the shepherd's crook in his hand and the cattle lumbering ahead of him, you might mistake Neil Heseltine for a rambler enjoying the breathtaking scenery of the Yorkshire Dales National Park. With wild blond-brown hair, khaki walking trousers, a blue fleece and neck buff, he was willing his cows to a fresh grassy hillside. A fourth-generation farmer, Heseltine farms 1,200 acres of rugged limestone uplands at Malham in North Yorkshire, a popular destination for walkers traversing the Pennine Way. His farm is a place of pasture and drystone walls over hills that rise and fall in lumpy undulation. On the lower land, the odd barn is connected by old stone lanes that have been used by monks, drovers and farmers for millennia.

Heseltine is one of a new breed of farmer dispensing with the wares of the feed merchant in favour of the grass beneath his feet. He grazes 150 Belted Galloway cattle – fantastic-looking animals with tufty black coats and a broad white band around their girth – and a similar number of sheep. His cattle stay out on the hills all year round and feed entirely on pasture. They have never been fed grain. His sheep herd is a different matter, having long been farmed 'conventionally', which means they were fed a lot of grain in the form of 'concentrates'.

In 2012, Heseltine realised that while his cattle grazing the grassy hillsides were profitable, his grain-fed Swaledale sheep were not. At best, they broke even. When he cut out the grain and halved the size of his flock, the environmental and financial benefits were dramatic. In 2012, his four hundred grain-fed sheep made him and his family just £478. Four years later, and with just two hundred ewes, his flock was making over £17,000 a year.[22] How was this possible? Because the cost of feed and veterinary drugs was largely removed, the animals could grow more naturally. Fewer sheep gave the pasture chance to recover and wildflowers to bloom and seed. The flock became healthier and less reliant on routine treatments of drugs. Heseltine had walked away from 'conventional' farming, cutting the umbilical cord that had kept the money flowing to animal feed and pharmaceutical companies. Instead, he was following a nature-friendly path that is often referred to as 'agroecology'.

Doing something different meant dispensing with years of industry advice and peer pressure. Breaking away from convention meant risking the wrath of those around him, whose response, as Heseltine noted, was often 'one of disapproval'.[23] His decision to go against the norm was helped by joining a group of peers in the Pasture-Fed Livestock Association, which at that time was a fledgling society set up by farmers who wanted to encourage the raising of ruminant animals on pasture and stop the 'grain-drain'.

Dr John Meadley, PFLA's founding chairman, told me about the group's motivations. 'Bucking a trend is not easy, particularly when many businesses benefit from selling feed, fertilisers and other chemicals to farmers, and who understandably feel threatened by a wholly pasture-fed approach that now does not need them. But change to an approach that seeks to nurture nature rather than to control it is inevitable, and we can already see it happening.'

Big business and the pressure of convention are major reasons why farming is struggling to respond to impending farmageddon. 'Economic pressures have encouraged farmers to increasingly feed their animals on high-energy commodities such as cereals and imported soybeans so that they grow faster and to keep cows indoors to avoid the energy "wasted" in walking to and from and around the fields', Meadley said. 'But we have shown that raising cattle and sheep wholly on pasture can generate benefits for the animals, for the environment and for the bottom line.'

Heseltine's example – and he is one of many – shows how separating farming from agribusiness can bring tangible benefits. Animals have better lives and farmers have better livelihoods. Though there's a good living to be made in nature-friendly farming, ensuring money goes to the right people is vital. In the battle for the countryside, I've learned that we shouldn't pity the 'poor' grain merchant.

POWERBROKERS

In the agricultural industry, power lies in a small number of hands; the economic and policy landscapes that created the current food system are reinforced by gigantic interests with great political sway.

To get a sense of how power is wielded, I spoke to Dominic Dyer, former CEO of the Crop Protection Association, an influential chemical industry lobby group.[24] Having previously worked for the Ministry of Agriculture and the Food and Drink Federation, Dyer, who is in his late forties, is a strong advocate for animal welfare.

It was 2008 when Dyer joined the agrochemical industry, representing companies including BASF, Bayer, Monsanto and Syngenta.[25] They wanted someone with a background and knowledge of the food industry to examine the use of agrochemicals in the food chain. He took on the role when the economy was in trouble and food prices were spiking, sparking a debate over global food production.

'Suddenly, there was an opportunity for the agrochemical industry to say, "Our technology is allowing us to grow more," and that's the sort of messaging that I pushed forward', Dyer admitted.

Though he was good at pushing the message, as time went on he began to question the integrity of his script, much of which was about the need to 'feed the world' and how agrochemicals would help do that while keeping prices low.

He recalled a time when the European Parliament was looking to restrict agrochemicals, which would mean big companies losing a lot of money. Dyer lined up the various industry bodies to warn of dire consequences for food prices, and the policymakers backed down.

'Once you had them all in one place, telling government "This is what's going to happen", ... they [the European Parliament] pulled it back', Dyer said. 'Looking back, some of those chemicals probably should have come off the approved list at that time but didn't because of what we did. It shows how government can be influenced by industry.'

By his reckoning, you could put twenty-five companies in a room and 'that's pretty much the food chain'. 'It taught me about the power of big industry', he said, 'about how power was in a relatively small number of people's hands.'

The same day as meeting Dyer, I'd been one of a hundred or so guests at Kew Gardens for Michael Gove's ministerial swansong as UK Secretary of State for Environment, Food and Rural Affairs. In a grand conservatory within one of London's great oases on a balmy day in July 2019, Gove cited 'agricultural intensification' alongside climate change and deforestation as a key driver of the demise of nature. It was an extraordinary speech, full of urgency and clarity – a clarion call for action that left many activists looking tame.

'Time is running out to make the difference we need, to repair the damage we as a species have done to the planet that we have plundered. Nature is everywhere in retreat – we have seen a catastrophic loss of biodiversity across the globe as a result of man's actions', Gove said. He went on to describe how British farmers would be rewarded for looking after the countryside after Brexit, something that EU agricultural subsidies had largely discouraged.

Globally, the public is providing $700 billion a year in global farm subsidies, more than $1 million per minute, much of which is driving the climate crisis and destroying wildlife. A detailed report in 2019 by the Food and Land Use Coalition showed that just 1 per cent of subsidies given to farmers are used to benefit the environment.[26] The report warned that without reform of these subsidies, humanity is at risk.[27]

Yet the need for change is not new. A decade earlier, four hundred scientists contributed to a groundbreaking report, *Agriculture at a Crossroads*, which called for the establishment of ecologically and socially sustainable agricultural systems to replace subsidies that encouraged industrial farming. The report, commissioned by the World Bank and six UN agencies with the endorsement of fifty-nine countries, showed the harmful impact of intensive farming on the world's poorest; despite increased global food supplies, poverty in sub-Saharan Africa, for example, has remained static. It called for an agroecological approach to feeding the world.[28]

In 2019, I had the privilege of appearing onstage with one of the report's lead authors, Professor Hans Herren, at the Oxford Literary Festival. A Swiss-born farmer and entomologist, Herren's early career took him to Nigeria, where he devised the world's biggest chemical-free agricultural pest-control programme. The project focused on fighting the cassava mealybug, the scourge of staple crops in Africa. Rather than fighting it with chemicals, he used a natural enemy – a South American parasitic wasp – and averted one of the world's biggest food crises, saving some 20 million lives.[29]

Herren has dedicated a lifetime to sustainable development and food production. A decade on from the report, I asked him how well we were doing at feeding the world with industrial agriculture. 'We produce way too much, and we waste way too much', was his reply. 'And all this has actually led to the world going past our planetary boundaries. We are on a planet with limited resources, so we can't always use more', he said.

Herren and his colleagues came up against the status quo, the narrative promoted by industrial agriculture that we need to produce more food to feed more people. This 'feed the world' narrative has become the standard-bearer for Big Ag, but it's one that Herren rejects. 'We produce double the food we need', he said. 'Which means that, even if we increase our population by two billion, from seven and a half, we still have plenty of food. So all this argumentation that we need more food, which always comes from the chemical industry, agribusiness and others, is basically wrong.'

Though Big Ag peddles fears that without intensification we'd all starve, the evidence paints a different picture: a wide-ranging global scientific analysis showed that organic farming actually increases yields in developing countries, where they are needed most. Compared with non-organic (intensive) farming, yields of animal and plant foods were on average 80 per cent higher. The scientists concluded that 'developing countries could increase their food security with organic agriculture'. And in developed countries, where so much is produced that vast quantities are either fed to farmed animals or thrown away, average yields were down by just 8 per cent – hardly the wholesale collapse the chemical industry would have us fear.[30]

Faced with a world of increasing agricultural intensification, Herren fears for the future. Intensification 'has led us to very low resilience in our agricultural system. We can't overcome droughts, nor can we overcome floods. And why? It's because our soils are no longer what they should be: a sponge that will take up the water when there's too much of it and release it when we need it', he said. His report referred to a 'vicious cycle'; current agricultural practice degrades the environment, causing farmers to move into marginal lands. A decade later, the UN FAO referred to the existence of 'a vicious downward spiral' in describing the effects on soil from industrial crop production; ultimately, they said, food production is 'seriously affected'.[31]

'We need to take a different pathway', Herren told me, 'farming in harmony with the environment.' In the face of mounting evidence of climate, nature and health emergencies, with food a big factor in each of them, I asked him why things weren't changing faster. 'Because we have a huge amount of power which is concentrated in big companies who actually make the politics of agriculture', he replied.

BATTLE LINES

Nairobi, and security gates clanked as I entered the epicentre of the battle for the planet. I had come to the African headquarters of the United Nations where the high-security entrance hall opened out into an expanse of lawns, meeting blocks and offices. A life-size figurine of an elephant stood sharp-eyed in the dappled shadow of surrounding trees. Although I had brought my camera with me, this was the nearest I would get to a safari.

In the warmth of a March morning in Kenya, I made my way through a winding avenue of nearly two hundred fluttering flags, one for each member country of the UN. As an animal protectionist, I was there not to represent a country but to fly the flag for humane food. At the UN Environment Assembly, the issues of climate change, fossil fuels and plastic were regularly on the agenda, but food had previously barely registered. On the lawn, a large, brightly coloured model yacht floated on a 'sea' of plastic bottles. Fascinated by the display, a young girl crouched down next to it, keen to get a better look. Her dad lifted her up and plonked her in the boat. It made me think that by the time the little girl reaches her father's age, it could well be food rather than plastic rocking the planetary boat.

Bringing together nearly 5,000 heads of state, ministers, business leaders, UN officials and civil society representatives, UNEA was a whirlwind of meetings and presentations tasked with tackling our environmental crises and steering humanity's interaction with the planet toward safety. It felt like groundbreaking stuff.

I had the privilege of participating in a session on healthy food for a sustainable planet, alongside a range of interested parties keen to discuss how we might deliver a decent future for the next generation. Not surprisingly, I found myself flanked by representatives from Big Ag who made the case for chemical fertilisers, pesticides, GMO and gene editing.

I listened as the representative from Croplife International talked about a Brazilian scheme to recycle the plastic drums that pesticides came in. The idea was clearly to ride the wave of concern about plastics, though the bigger case, surely, was not the plastic drums but the chemicals that they contained. After all, the overuse of chemical pesticides has led to a situation in Brazil where one person is poisoned by them every ninety minutes.

I used my keynote speech to give a stark warning about the environmental consequences of our current food system, sketching out the farmageddon scenario that was leading us towards a major planetary crisis. At the heart of all these threats, I said, is industrial agriculture: the use of pesticides, fertilisers, cages and monoculture crops to produce 'cheap' meat and milk. I spoke about the need for a sustainable vision for food, farming and nature. I pleaded for consensus among leaders in government and business, a global agreement for a regenerative food system without factory farming or excessive meat consumption.

I wasn't alone; other speakers talked about the importance of reducing meat and dairy production, in order to stabilise the climate and save nature.

I was heartened by the level of support I received, yet one incident which happened during the plenary feedback session to world governments would underline the scale of the challenges ahead. The UN plenary had already heard from child ambassadors, representing the voice of the young, who had commented on what they saw as important: plastics, the loss of wildlife and food. The message was powerful: in protecting the planet, grown-ups had failed. They were clear that to get things done, grown-ups had to stop arguing like children.

When it came time for moderators from the grown-up workshops to report back to UN and government officials, the representative of the fertiliser industry took to the floor to summarise 'my' session; I was shocked to find that my keynote message, along with those of others who felt similarly about the need for food system transformation had been shamelessly airbrushed out. Instead, the view that was heard by governments focused on the role of fertilisers and gene editing in delivering food for the future. My call to reduce meat consumption and shift away from Big Ag was glossed over. Instead, what the world leaders heard was how some communities in developing regions needed to eat *more* meat. I was astonished by how the feedback misrepresented our interventions. The veteran environmentalist and former WWF policy lead Duncan Williamson agreed with me. As he said, 'It felt like the script had been written before the event and that it didn't matter what actually happened on the day.'

UN officials and governments had been given a thoroughly incomplete picture on which to base decisions affecting the future of our planet. For now, it seemed the benefits of regenerative farming and balanced diets would remain hidden for a bit longer. It felt like battle lines had been drawn: Big Ag wasn't letting go without a fight.

12

PANDEMICS ON OUR PLATE

PROTECTING PEOPLE MEANS PROTECTING ANIMALS TOO

Wartime. That's how it felt. Entire countries were confined to their homes for their own safety. Governments took stringent measures against an invisible enemy: Covid-19. Within a dozen weeks in 2020, a viral outbreak in the remote Chinese city of Wuhan had a profound effect on the lives of people all around the world. In January, grainy television pictures from China showed desperate measures being taken to contain the virus. Streets, buildings and hospitals were deep-cleaned by people wearing protective suits, while doctors were at their wits' end. The world watched in disbelief.

Like an avalanche gathering momentum and sweeping all before it, this new strain of coronavirus swept across the planet. Aided by our mobile society with copious international travel and global trade, the epicentre of the outbreak quickly moved from China to Europe. Italy and Spain bore the brunt. The European Parliament shut down, while in America, President Trump warned of 'painful' times ahead.[1]

Countries dealt with coronavirus differently. In Colombia, people were allowed to leave the house on specific days according to their ID number. In Serbia, a designated dog-walking hour was introduced. In Belarus, the president recommended vodka and saunas as a way of avoiding infection. More commonly, governments issued strict social distancing rules and restricted non-essential travel.[2] As country after country went into lockdown, normal life ground to a halt. Sports venues, theatres, cinemas and shops were closed, and even parks and nature reserves suddenly became out of bounds. Travel bans grounded many airline flights as countries closed their borders, and travel became frowned upon; cars stayed on drives, with roads as empty as Christmas Day.

Governments focused on preventing health services being overwhelmed, while nations celebrated those involved in the front line – doctors, nurses and other key workers desperately trying to keep everyone well. By April 2020, with half the world's population in lockdown and the global death toll reaching 30,000, the UN's secretary general, Antonio Guterres, described it as humanity's worst crisis since the Second World War. The disease, he said, 'represents a threat to everybody in the world and … an economic impact that will bring a recession that probably has no parallel in the recent past'.[3]

World leaders talked of being on a 'war footing',[4] while the UK health secretary, Matt Hancock, described it as a 'war against an invisible killer'.[5] In the space of a month, Britain went from being reassured by the Mayor of London that there was 'no risk' of catching coronavirus on the city's Underground[6] to a situation where police were given the power to disband gatherings of more than two people. In the same month, Hancock and the prime minister, Boris Johnson, both tested positive for Covid-19.[7]

As in wartime, governments armed themselves with emergency powers and took swift and extraordinary measures, in a bid to defeat the unseen enemy. 'Non-essential' business was shut down and life was effectively put on hold. As a consequence, economies took a nosedive, stock markets plunged and city centres fell silent. Leaders struggled as the situation spiralled out of control.

People in developing countries were hit particularly hard, and there were scenes of desperation and unrest. In Tunisia, several hundred people protested against a lockdown that disproportionately hit the poor: 'Never mind coronavirus, we're going to die anyway! Let us work!' shouted one protester at a demonstration on the outskirts of Tunis. Africa's biggest city, Lagos in Nigeria, faced the daunting challenge of enforcing a lockdown on millions of people living hand to mouth in some of the world's biggest slums.[8]

PANDEMIC ON A PLATE

While the emergence of Covid-19 has been linked to human consumption of wildlife, the resulting pandemic has a lot to say about the future of food and farming. The outbreak was first linked to a market in Wuhan[9] that sold seafood and live animals, including live wolf pups, bamboo rats, squirrels, foxes and civets.[10] Although bats are thought to have been the original source, scientists suspect that the virus may have jumped to humans via an intermediary, possibly a pangolin, a scaly mammal that is said to be the most trafficked mammal in the world. In 2019, an estimated 195,000 pangolins were trafficked for their scales.

Scientists have long warned of the dangers posed by the illegal trade in wildlife. After the outbreak of SARS – severe acute respiratory syndrome – in 2002–3, Chinese scientists conducted research into the risks of trading and eating wild meat. Zhang Jinshuo of the Institute of Zoology at the Chinese Academy of Sciences, who took part in the investigations into the source of SARS, said: 'We later published many papers and popular science articles, urging everyone to stop eating wild animals and not to have too close contact with wild animals. Only the health of wild animals and the health of ecosystems can [secure] human health.'[11]

Reports of the conditions in which the animals in these wildlife markets were kept made for grim reading. As well as moving to close down the illegal trade in wildlife, nearly 20,000 wildlife farms that raised species including peacocks, civets, porcupines, ostriches, wild geese and boar have been shut down across China in the wake of the coronavirus. Only a few years earlier, wildlife farming was promoted by government agencies as a way for rural Chinese people to get rich. And just weeks before the coronavirus outbreak, China's State Forestry and Grassland Administration was actively encouraging citizens to farm wildlife such as civets, a species identified as a carrier of SARS.

Covid-19 caused a massive rethink. 'The coronavirus epidemic is swiftly pushing China to re-evaluate its relationship with wildlife', Steve Blake, chief representative of WildAid in Beijing, told the *Guardian*. 'There is a high level of risk from this scale of breeding operations both to human health and to the impacts on populations of these animals in the wild.'[12]

There was increasing concern that fears over the origins of the new coronavirus could spark a backlash against animals in the wild. That it might set off a new wave of ecophobia: a fear of nature. There were suggestions of the 'need' for the 'ecological killing' of disease-transmitting wild animals such as pangolins, hedgehogs, bats, snakes and some insects, with some authorities poised to cull wild animals as a result.[13] However, as the world came to realise that the wildlife trade could have sparked a pandemic, we also began to regard agriculture as having a big bearing on the prospect of future disease outbreaks.

STRONG PARALLELS

Although it is widely thought to have originated in wildlife, Covid-19 shows parallels with other viruses that have stemmed from industrial animal farming. For decades, scientists have been predicting that a pandemic would originate from farm animals. Swine flu and highly pathogenic strains of avian influenza, which originated in pigs and chickens, are believed to come from keeping sentient creatures in caged, cramped and confined conditions that provide the perfect breeding ground for novel strains of disease.

The H5N1 virus, a highly pathogenic strain of bird flu, emerged at a time when the poultry industry in the Far East was expanding massively. It was first identified in Hong Kong's live bird markets and chicken farms in 1997, where it caused six human deaths. From 2003 it spread across East Asia, at exactly the point when poultry production was becoming more intensive. China reared three times as many chickens in 2005, when bird flu was rampant, as in 1990.[14]

The H5N1 virus spread across Asia, the Middle East, Europe and Africa. It has been found on chicken, goose and turkey farms, and in wild birds, mainly swans and geese. By August 2011, 564 people were confirmed to have been infected, of whom 330 died – a fatality rate of almost 59 per cent.[15]

H5N1 is not easily transmitted between people, but scientists have suggested that just a few mutations would allow it to become as infectious as seasonal flu. An editorial in *New Scientist* described the risk of a pandemic as 'fact, not fiction'.[16] The fact that flu viruses can spread between pigs, humans and birds has in recent times caused one deadly pandemic and promises the nightmare prospect that there will be more.

High on a desert plain in south-east Mexico, in 2011, I visited the epicentre of one such pandemic: swine flu. Small-town streets were adorned with graffiti providing public health information. A painted wall, the colour long faded, had been decorated with a cartoon figure dutifully washing his hands alongside a message emphasising the importance of personal hygiene. I stared at it with detached surprise, but since Covid-19, the message has seemed much closer to home.

I was in La Gloria, a community of little more than 2,000 people that had the unenviable reputation for being 'ground zero' for swine flu (H1N1), the respiratory disease that swept across the planet in 2009 and 2010.[17] The first reported case was identified there in mid-February 2009.[18] This new and highly contagious virus contained genetic material from a mixture of pig, bird and human influenza, a combination that had never been seen before. The disease strain, which was found to share genetic material related to both North American and European pigs,[19] spread rapidly from country to country, a new type of flu virus that few people were immune to.[20]

I arrived in Mexico a year after the pandemic. The media spotlight had long moved on, leaving the residents of La Gloria alone with a network of industrial pig farms owned by Granjas Carroll de Mexico, the country's biggest pig producer.[21] Its facilities are scattered across a flat, arid plain of dry grasslands studded with cactuses. I can still feel the shooting pain I felt when I stubbed my boot-clad foot on them, their thorns like needles piercing my flesh. I walked through neatly furrowed fields populated with stacks of corn arranged in pyramids. From a distance, they resembled a choreographed crowd looking at the mountain range that bordered the valley. The metallic outline of a pig factory farm was visible through this corn congregation, stark warehouses hugging the ground in close rows. There was no sign of any pigs; they were kept inside, their presence betrayed only by the odd muffled squeal and the searing stench of swimming-pool-sized lagoons of liquid faeces. Even the most imaginative commentator would be hard-pressed to describe these as farms. They looked more like military installations, with each rectangular building flanked by a glinting feed silo the shape of a rocket launcher. For the uninvited, there was no way in; the perimeter was secured by impenetrable fencing. And for the four-legged inhabitants, there was no way out.

Whether the locals wore steep-brimmed cowboy hats or colourful baseball caps, swine flu had left an indelible mark on them, as had the pig farms. I met people who were worried for their children, their livelihoods and their way of life, and the pig farms that had spread across the valley were the subject of their ire. I heard how promises of jobs had failed to materialise, with the pig farms instead bringing nothing but flies, smell and pollution. Guadalupe Gaspar, a sixty-eight-year-old farmer and a leader of local protests against the pig farms, was filled with anger. 'We are living in a time bomb', he told me. 'We don't know when something else bad is going to happen to us. The government must get rid of the farms because while they remain, the pollution will continue, and I am sure there will be more new diseases.'

The farms started arriving in the area some years before swine flu. Soon after the first farm was opened, villagers reported noticing a change in the quality of their groundwater. They began writing letters of complaint to the federal government, the state authorities and the town council. When the next farm was built, protests ensued and roads were blocked. But nothing changed, and swine flu arrived two years later. A young boy in La Gloria, Edgar Hernandez, was the first confirmed case. In recognition of the town's claim to fame, a bizarre four-foot statue of him was erected in a local park. La Gloria seemed strangely keen to remember this painful chapter in its history.

What was clear to me from my visit to La Gloria was that, whether or not swine flu had originated in the local pig farms, the rest of the world had moved on and the salutary lesson about the risk posed by factory farms had been forgotten. According to the US Centers for Disease Control, swine flu was within a year responsible for more than 12,000 deaths in the US and an estimated 151,700 to 575,400 people worldwide.[22] It was a warning to the world. But once the immediate panic was over, most people stopped worrying about catching swine flu, just as they had stopped worrying about bird flu.

With respect to Covid-19, the prospect that we will fail to learn the lessons of the current pandemic is a major concern. In the wake of previous disease outbreaks, we have been quick to forget. There has also been a tendency to cling to false solutions. When faced with a disease that involves farmed animals, the industry's reaction is to lock livestock inside. After all, if they are confined indoors in 'biosecure' units, they are surely protected. However, such logic overlooks the fact that these intensive farm buildings provide the perfect breeding ground for new and more dangerous strains of disease. Keeping too many animals in too small a space, often in dark, filthy and overcrowded conditions, provides the conditions a virus like avian influenza needs to spread rapidly. As it goes through the flock, replicating wildly, differences can occur in the virus's DNA, which can cause new – and potentially more deadly – strains to emerge. Contrary to what the farming industry often leads us to believe, keeping farmed animals indoors *increases* the risk of disease.

BLACK SWAN

Unlike the swine flu pandemic, Covid-19 is thought to have its roots in wildlife. There are hundreds of coronaviruses, most of which circulate among animals including pigs, camels and cats. Bats are another common carrier, but they are unlikely to have transmitted the virus to humans directly – as with most similar viruses such as SARS and MERS, an intermediary animal is thought to have been responsible.[23] Civets were thought to be intermediary for SARS and camels bridged the human-animal gap for MERS. It has been suggested that pangolins were Covid-19's intermediary; while Chinese legislation sentences anyone caught selling them to ten years in prison, they remain the victim of a massive trade in illegal trafficking.[24]

When SARS emerged in 2002, more than 8,000 people fell ill and 774 died across the globe, but Covid-19 surpassed these numbers within two months.[25] As the 2020 coronavirus pandemic sent shockwaves around the world, some started to recognise that the long-term problem behind the current crisis was the collapse of nature. With Europe deep in pandemic, Germany's federal environment minister, Svenja Schulze, released a statement saying, 'Science tells us that the destruction of ecosystems makes disease outbreaks including pandemics more likely. This indicates that the destruction of nature is the underlying crisis behind the coronavirus crisis. Conversely, this means that good nature conservation policy that protects our diverse ecosystems is a vital preventive health care measure against new diseases.'[26]

New and devastating disease outbreaks have become known as 'black swan' events.[27] Could it be that this one, or the next, has its roots in the demise of the natural world caused by industrial agriculture?

OVERWHELMING NATURE

As humanity continues to destroy the natural world, felling tropical rainforests and erasing pristine habitats, so we come into contact with new species of life, including viruses, heightening the risk of new black swan events.

As David Quammen, author of *Spillover: Animal Infections and the Next Pandemic*, wrote, 'We invade tropical forests and other wild landscapes, which harbour so many species of animals and plants – and within those creatures, so many unknown viruses. We cut the trees; we kill the animals or cage them and send them to markets. We disrupt ecosystems, and we shake viruses loose from their natural hosts. When that happens, they need a new host. Often, we are it.'[28]

A major reason for our encroachment on nature is our desire for cheap meat. If you take away the deserts, the ice caps, the mountains – in food terms, the land that is unusable – nearly half of what's left is used to grow food. Four-fifths of that is devoted to producing meat and dairy. The rising demand for animal protein has fuelled agricultural expansion, pushing ever deeper into marginal lands and wild spaces.

Much of the world's arable land is now shared with factory-farmed animals, with 40 per cent or more of our grain harvest being fed to animals that are caged or confined. Globally, that's an area of cropland equivalent to the entire European Union or half the United States of America. Growing food – or 'feed' – using heavy applications of chemical fertilisers and pesticides, means that nature is wiped away.

People tend to associate deforestation with logging, or with felling trees to make way for housing and crops for human consumption; in fact, the real driver is the farming of feed crops like soya and corn. As I discovered during trips to Brazil and Argentina, vast areas of rainforest and savannah are turned over to these industries. Grazing animals, and especially cattle, are blamed for the destruction of the rainforest, yet in Brazil I saw that the global demand for 'cheap' meat is the real driver. Long-standing cattle pastures on the country's savannah plains are being ploughed up for feed crops. The demand pushes up land prices, allowing cattle farmers to buy land deeper in the forest, thus accelerating rainforest destruction in what is known as a 'land use cascade'.

It is because so much of the world's harvest is squandered in this way that we encroach on more forests, which brings us into contact with a new array of wild animals, plants and viruses. We thus put ourselves at risk of a fresh pandemic every day, while also loading the dice of sustainability against ourselves. We are literally encroaching on our own future.

Soya, maize and palm are the major drivers of deforestation and habitat loss around the world. Thus, two of the major problems with factory farming become apparent. The first is the fact that farmed animals are kept in conditions that provide the perfect breeding ground for novel viruses. The second is the increased footprint of the animal feed industry, which sweeps into forests, disrupting wildlife communities and potentially bringing people into contact with new viral 'invisible enemies'.

'Never before have so many opportunities existed for pathogens to pass from wild and domestic animals to people', said Inger Andersen, executive director of the UN Environment Programme. 'There are too many pressures at the same time on our natural systems and something has to give. We are intimately interconnected with nature, whether we like it or not. If we don't take care of nature, we can't take care of ourselves. And as we hurtle towards a population of 10 billion people on this planet, we need to go into this future armed with nature as our strongest ally.'[29]

RUSSIAN ROULETTE

The coronavirus pandemic and swine flu a decade earlier have shown how treating animals – whether they are domesticated or wild – as commodities is like playing Russian roulette with people's health. I've come to see that showing more humanity to animals today is key to reducing the risk of devastating diseases tomorrow. Respecting the sentience of animals – their ability to feel pain, to suffer and to experience joy – lies at the heart of future disease-control strategies. In recognition of this, policymakers are increasingly using the phrase 'One Health'. By this they mean that the health of people depends on the wellbeing of animals and a thriving environment. When we fail to respect animal welfare or protect nature, we undermine our own health and viability.

The coronavirus epidemic was not just a warning, but a demonstration of what is going wrong and what life could become. Where our global lifestyle once seemed invincible, suddenly it seems fragile. It begs the question, might our consumptive lifestyle be our Achilles heel?

ENVIRONMENTAL LESSONS

Nationwide lockdowns during the coronavirus pandemic resulted in rapid decreases in air pollution, giving the environment time to breathe. Within weeks, major UK cities like London, Birmingham and Cardiff experienced drops in air pollution of up to a half,[30] and the same phenomenon was seen around the world. In what was described as the 'largest-scale experiment ever' in terms of industrial emissions, satellite imagery from the European Space Agency showed drastic reductions in China and Italy. Nitrogen dioxide levels fell due to there being fewer cars on the roads and fewer emissions from industrial power plants. Changes in the atmosphere over northern Italy were particularly striking. Smoke from a dense cluster of factories tends to get trapped against the Alps at the end of the Po Valley, but within a fortnight of lockdown, nitrogen dioxide levels in Milan and other parts of northern Italy had fallen by about 40 per cent.[31]

Air traffic, a major source of greenhouse gas emissions, fell sharply in the wake of coronavirus. Within three months of the first cases being recorded, the airline industry was predicted to lose more than $100 billion in sales.[32] While the business impact was catastrophic, the environmental benefit was clear. Covid-19 was driving the kind of reductions in emissions that neither governments nor industry were willing to make happen. As Paul Monks, professor of atmospheric chemistry and earth observation science at the University of Leicester, told the *Guardian*, 'We are now, inadvertently, conducting the largest-scale experiment ever seen. Are we looking at what we might see in the future if we can move to a low-carbon economy? Not to denigrate the loss of life, but this might give us some hope from something terrible. To see what can be achieved.'[33]

While no one wanted such a terrible pandemic, which has claimed so many lives and devastated countless others, this dark cloud might have a silver lining that may save us from worse things to come. Living through the unimaginable, a shared global experience of adversity, could provide people and policymakers with the necessary insight to rethink how to tackle mounting problems. If left unaddressed, they could well leave a diminished planet for future generations.

STARK CONTRASTS

Governmental action on Covid-19 stands in stark contrast to efforts to tackle factory farming, climate change and the collapse of nature. The scientific evidence clearly shows that the level of threat posed by unsustainable farming practices and climate change could plunge the world into unimaginable crisis, but few leaders are willing to do what it takes to prevent it. As Jean-Claude Juncker, president of the European Union, said of climate change, 'We all know what to do, we just don't know how to get re-elected once we've done it.'[34] Businesses locked into achieving profit for shareholders are unwilling to turn their world upside down, so we are locked in a circle of short-term thinking. Anything other than incremental change seems radical and unrealistic.

Yet the pandemic has shown that for the sake of a decent tomorrow, drastic measures are needed today. If we don't take urgent action to address the food system, climate and the collapse of nature, we could again find ourselves battling an enemy of our own making in a war without end.

As Karl Falkenberg, a former European Commission director general for the environment, pointed out, society rarely makes major changes without first taking a hit. Talking about food, climate and the environment, Falkenberg told a London conference, 'Why, with all the knowledge that we have, can we still not get the right governance decisions? Why do we continuously do the wrong things that we know are wrong until the next disaster hits us? We do need bloody noses before collectively we start modifying systems.'[35] And with Covid-19, we got one hell of a bloody nose.

There is still time to solve the climate crisis and the decline of nature. Although things are already bad, we are yet to reach crisis point, at least one that hits the heart of society. As for the world's oceans, they are likely to have commercially catchable fish in them for a few decades more, and the disappearance of the world's soils feels like a lifetime away. However, as Falkenberg put it, 'Collectively, we are living beyond the supporting capability of this blue planet. And we are doing it while we know that this is the only planet we have. It is a fundamental contradiction. It's short-termism in the worst form we can imagine.'

From the point of view of politicians elected to govern for a four- or five-year term or so, big problems can be left to someone else to sort out. And for business, the future depends on the next quarter's results, so there is a tendency to leave it to consumers to decide through their buying choices. However, thanks to pricing signals that make 'bad' stuff cheap and good stuff hard to come by, and labels that make it hard to decipher which is which, consumers are essentially blindfolded. In Europe, about 80 per cent of citizens say they want to buy fairtrade, organic and sustainably produced food, but only 5 to 6 per cent actually do.

With Covid-19, the immediacy of the 'attack' meant that political leaders did things for the common good that would previously have been unimaginable. The devastation from this disease wasn't something politicians could put off for another day or leave for their successors to pick up. No, this was *their* problem, and *their* jobs depended on how they acted. Businesses were given no option but to comply. Electorates were under serious threat, so the elected had no choice but to act.

Perhaps the response to Covid-19 will establish a new way of looking at threats beyond the immediate horizon. Perhaps, as a global society, we will now find it within ourselves to take the future seriously. The response to Covid-19 has set the template for what could be done to stave off other threats facing society. And while this pandemic has been linked to wildlife in a wet market, the next one could come from an incarcerated pig or chicken, animals reared like commodities and fed from the ashes of deforestation. Either way, factory farms could be making victims of us all. Never before has there been a more potent example of how the health of animals and people are so closely interlinked. In the war against invisible enemies, protecting people means protecting animals, too.

13

BUSINESS WITH BOUNDARIES

HOW CORPORATES WILL ADAPT OR DIE

Though we rely on it for our livelihoods, our pensions and our social infrastructure, business is often blamed for the environmental crisis. The focus of the corporate community on making a profit has long put it at odds with animal welfare and environmentalists; using the Earth's resources to produce goods or services has all too often been about short-term returns at the expense of long-term sustainability.

If Covid-19 has taught us anything, it's that business is dependent on our planet being stable – after all, there are some things money can't buy. As the former Bank of England governor Mark Carney argued, society is in danger of embodying Oscar Wilde's old aphorism: 'knowing the price of everything but the value of nothing'.

By the third decade of the twenty-first century, the business world is starting to wake up to the fact that our economy has to work within planetary boundaries; the illusion of infinite growth in a world with finite resources is starting to wear thin. If business is to stay in business, it will need to adapt to survive.

Norwegian physician Dr Gunhild Stordalen is at the forefront of those who are making the corporate world aware of our planet's limitations. Prior to her separation from billionaire businessman husband, Petter A. Stordalen, the couple were known as the 'Bill and Melinda Gates of Norway' for their combination of business and philanthropy.[1] Stordalen is the founder and executive chair of the EAT Foundation, a global platform that links climate, health and sustainability issues to transform the global food system. After sharing a stage with her at a conference on sustainability in London organised by the *Financial Times*, I asked her what role she sees business playing in getting us out of the mess we're in.

'The businesses which will be successful are those who will provide goods and services that consumers want, and that can demonstrate that their goods and services have positive impacts on people and planet', Stordalen said. She sees business shifting quickly in the coming decades towards climate-neutral, regenerative production and circular business models. It is a vision that has a great deal in common with my thoughts about how we should change our food system.

Regenerative production is based on growing renewable resources using agricultural methods that eschew the need for inputs like chemical fertilisers or pesticides, instead using natural cycles to grow food while replenishing soil, water quality and natural services like insect pollination. Circular consumption would cut out waste, thereby drastically reducing our use of resources. And circular economics would involve replacing the current 'take, make and throw it away' model of using resources with a transition to renewable energy sources, recycling and eliminating waste.

'Innovators, those that see opportunity in regenerative, nature-positive, health-promoting and socially conscious business models, are likely to be the most successful. We're seeing this already in Europe, with parts of Asia, Africa and the Americas following suit,' Stordalen said.

Having worked closely with Professor Johan Rockström of Stockholm University, who came up with the concept of a 'safe operating space for humanity', she is only too aware that we are starting to reach the limits of planetary boundaries. Her foundation co-commissioned what has become known as the 'EAT-Lancet planetary health diet', a peer-reviewed report involving many scientists, which calls for a reduction in our consumption of meat and other animal-sourced foods if we are to avoid global meltdown. It recommends a per 'capita consumption of no more than an average of 300 grams of red meat and poultry and 200 grams of fish per week in a diet that is both healthy and environmentally sustainable'.[2,3] This would mean halving our global meat consumption, with the UK and EU cutting by two-thirds or more, and the US by four-fifths, to avoid exceeding planetary boundaries.

Rockström's team of scientists at the Stockholm Resilience Centre identified nine processes with planetary boundaries, including biodiversity, climate, nitrogen and phosphorus flows, ocean acidification and the conversion of wildland habitats for human use.[4] According to their analysis, humanity has already breached the boundaries of biodiversity and nitrogen flows, with climate and land-use change not far behind.[5] Industrial agriculture and increasing meat consumption are major contributors to all those danger areas. Converting wildlands into fields to feed industrially reared animals or ploughing up pastures and moving cattle deeper into forests are taking a toll on the world's wildlife.

One association with industrial agriculture that often goes unnoticed is nitrogen pollution from artificial fertilisers and slurry. Nitrogen and phosphorus are essential for plant growth and are provided for in nature-friendly farming by the manure from free-roaming animals. Yet in industrial agriculture, producing and applying artificial fertilisers is a major cause of pollution. I've seen this in many places, not least in the Midwest of America, where corn is grown using fertiliser made in chemical plants as big as industrial estates, much of which washes into waterways that feed into the Mississippi. It ends up in the Gulf of Mexico, where it is causing one of the largest marine 'dead zones' in the world. There are already hundreds of such areas in the world, and the number is doubling every decade.

Food is a big reason why humanity is going beyond our safe 'operating space'. At Stordalen's EAT Stockholm Food Forum in 2019, billed as the world's leading platform for global food transformation, Rockström declared that we are 'failing' to address the problems facing humanity, with our food system the primary culprit *and* victim.

Recent intergovernmental reports on climate and biodiversity combine to deliver a stark warning on what breaking planetary boundaries means. As Rockström said, 'we have the scientific evidence for a planetary emergency'.[6] Continuing as we are, he warned, will take us 'irreversibly towards a world that can no longer support humanity and food systems'.

Human impact on the environment is reaching a planetary scale to the extent that scientists talk about a new epoch, the Anthropocene: humans are now the single most important factor in environmental change, a force that will be permanently imprinted in the geological record. It is against this backdrop that Stordalen sees the urgent need for business to respond. 'For business, the planetary boundaries provide a solid and quantifiable risk measure, and we should be doing everything we can to ensure that economic activities not only remain below these boundaries, but that they become positive forces in regenerating and restoring earth system processes', she said.

Government will be crucial in getting business to play ball. I asked the journalist George Monbiot for his view on how the two will interact. 'Well, so much depends on whether government is prepared to regulate it', he said.

Monbiot sees two possible paths. One is that government keeps procrastinating and failing to grasp the nettle and 'as a result, business just carries on as usual until we all go to hell in a handcart. Which, I have to say, unfortunately is quite a realistic scenario. The other possibility is that suddenly, prompted by mass youth strikes and people waking up enough to change their governments and demand governments which are going to protect us from disaster, we see a completely new regulatory environment where you cannot do business, you will not have a licence to operate, unless you are doing so within ecological constraints.'

Monbiot suggests that business will have a big role in driving technological innovation like alternative proteins from non-animal sources such as cultured meat, but also in shifting the way the world views economic growth. As he says, 'We're also going to need to move to a no-growth economy because, even with substitution, a level of economic activity beyond a certain extent is going to be disastrous. Especially if it's passed planetary boundaries.'

In the energy sector, the challenge of climate change has led business leaders and governments to unite in driving the 'global energy transition' towards net-zero emissions by the middle of the century.[7] Without such a drive, emissions targets set by the Paris Agreement will be unobtainable. Energy accounts for two-thirds of global emissions. The year 2020 was supposed to be a turning point in the transition, until the Covid-19 pandemic threatened to put gains on the backburner. However, it also showed how the world can do things differently in the face of an existential threat. The skies in some of the most polluted areas of the world cleared, and so too did the links between a successful energy transition and long-term economic prosperity.[8]

Without reform, food, which is responsible for more than a quarter of greenhouse gas emissions,[9] could also put the targets of the Paris Agreement out of reach. Industrial meat and dairy do the most damage to climate, nature and human health, which is why I call for a similar global agreement to transition away from industrial agriculture and meat-heavy diets.

The impact of the food industry on future sustainability has prompted a senior economist at the International Monetary Fund, Nicoletta Batini, to call for a 'Great Food Transformation'. While the climate implications of fossil fuels have received much attention, recent research by the UN's Intergovernmental Panel on Climate Change shows that what we eat, how we produce it and how it gets to us exerts an even greater impact on the global environment and public health.

Writing in *Finance and Development*, Batini argued that greening food production and managing demand are crucial if we are to meet the UN's 2030 Agenda for Sustainable Development and the environmental pledge behind the Paris Agreement.[10]

An Italian economist based in Washington DC, Batini is known for thinking innovatively about monetary policy and is also passionate about food. A mother of twin girls, she is motivated by wanting to leave a decent planet for them and espouses respect for life in all its forms.[11] She has become recognised as the IMF's expert on food systems and sustainability, building a high-level network of people calling for a transition to the sustainable production and consumption of food.

In conversation with Batini and from reading her writing, I learned that she believes in change through targeted economic, financial and trade policies, as well as structural reforms. 'Currently, in many countries, large amounts of taxpayers' money are spent on subsidies that encourage otherwise unprofitable, unsustainable meat and dairy production predicated on the systematic inhumane treatment of farmed animals, as well as growing monoculture commodity crops for animal feed', she wrote. 'These subsidies should be redirected towards sustainable farms producing plant-based protein for human consumption and towards incentives for innovation on alternative proteins and smart farming technologies.'

Batini sees providing financial assistance as essential in the move to good food and farming practices. 'Replacing production subsidies with ecological payments to sustainable farmers could reorient industrial agriculture, adding to the climate mitigation potential, while reducing negative impacts on farm incomes.'

Measures to foster 'aggressive conservation' might include land tenure legislation and financial incentives to encourage landowners to protect ecosystems, especially in regions that host the planet's rainforests. At an international level, she has argued for the establishment of a fund to compensate countries that forgo trade in commodities where production might threaten critical ecosystems.

The tax system could also help power the transition away from diets that are heavy in industrially produced meat. Just as carbon taxes aim to reduce the carbon footprint of the energy sector, taxes could be levied on 'unsustainable' and 'unhealthy' foods to bring consumption in line with nutritional recommendations.

'The average US retail price of a Big Mac, for example, is around $5.60. But with all the hidden expenses of meat production (including health care, subsidies and environmental losses), the full burden on society is a hefty $12 per sandwich – a price that, if actually charged, could more than halve the US demand for burgers', Batini wrote, citing David Robinson Simon's *Meatonomics*. Tax revenue could be used to offset the cost of healthy, sustainable foods, transforming the way we farm and turning the food system the right way up, providing healthy food, affordable for all. On a similar theme, Olivier De Schutter, former UN Special Rapporteur on the right to food, has stressed that 'any society where a healthy diet is more expensive than an unhealthy diet is a society that must mend its price system'.[12]

Batini believes that it is hard to overstate the planetary benefits of greening food and farming. She is convinced by the interconnectedness of food, human and animal well-being and nature. 'If we can muster the will before it is too late, we can have our nutritious food, thriving economies and a habitable planet, too.'[13]

I asked her how economics needed to evolve, and she told me about an IMF-hosted conversation with Sir David Attenborough in which he argued that we cannot hope to sustain life without taking care of nature. A summary of the conversation described how sustainable prosperity and nature are intertwined, using a comparison with the financial world to make the point: 'we would not eat into capital to the point of depletion because that would bring about financial ruin. Yet in the natural world, we have done this repeatedly with fish stocks and forests, among many other resources – in some cases to the point of decimation. We must treat the natural world as we would the economic world – protecting natural capital so that it can continue to provide benefits well into the future.'[14]

This interconnectedness is described by some people as a 'doughnut' of societal and planetary boundaries. Our current economics model focuses on growth and gross domestic product, but simply adding up the monetary cost of everything ignores the need for growth that avoids damaging natural resources. It also fails to take into account whether endless economic growth is really meeting the needs of people. Growth continues, but more than 820 million people remain hungry worldwide.

'Doughnut economics' measures the performance of an economy by the extent to which the needs of people are met without damaging the planet. First published in 2012 in a report for Oxfam by Kate Raworth, the concept rapidly gained recognition internationally, including at the General Assembly of the UN.[15]

But whether within the existing economic model or a new one, many people are now recognising Batini's call for a 'Great Food Transformation'.

My take is that the key to transformation lies in a global agreement at UN level to move away from industrial agriculture towards planet-healthy diets, something that would rely on three main points:

1. A move away from industrial agriculture towards regenerative production; a shift away from monoculture crops and confined animals and towards nature-friendly farming practices could be a real win-win. Farmed animals would live better lives, and the countryside could breathe again. Wildlife would be restored, and regenerative soil practices could help tackle climate change. Using rotational farming methods, planting cover crops and perennials and eliminating monocultures could also lock as much as sixty tons of carbon per acre in the soil, reducing the carbon dioxide in the atmosphere. Rattan Lal, a leading soil expert, has calculated that 'a mere 2 per cent increase in the carbon content of the planet's soils could offset 100 per cent of all greenhouse gas emissions'.[16]

2. A reduction in meat and dairy production by at least half globally in the next thirty years, with high-consuming parts of the world needing to reduce consumption further and faster: the EU and UK by two-thirds or more, and the US by four-fifths. Reductions should be focused on the intensive sector: feedlot beef and factory-farmed chicken, pigs and fish. We should avoid the overly simplistic impulse to cut grazing red-meat animals in favour of chickens and pigs, which would only serve to confine

more animals to lives of misery on factory farms. We should promote a circular consumption model where food waste is reduced to a minimum and properly treated, before being recycled as feed for pigs and poultry.

3. Restoring nature through better land-use strategies, including rewilding and forest planting, and ending deforestation. We should protect the lungs of the Earth and revitalise the natural world, humanity's life-support system. Grazing animals should be on permanent pastures and integrated with crops on mixed farms. We should avoid farming animals in the wrong place, such as in the Amazon rainforest or on uplands that should be covered by trees.

A sustainable future is increasingly reliant on creating a global food transition. Corporations, governments, civil society and the UN need to work together on changes that are vast enough to save the planet and all living beings.

14

SAVING OURSELVES

FROM EXTINCTION TO RESTORATION

Deep among forest-clad slopes of Rwanda's Virunga Mountains, a young David Attenborough was creating a moment that would become one of the most iconic in wildlife history as he met a family of endangered mountain gorillas. Surrounded by dense vegetation, he lay among several powerful apes, his sense of joy palpable. Looking a tad overdressed, this quintessentially English gentleman found his composure melting away as a mischievous young gorilla called Poppy tried to remove his shoes. 'There is more meaning and mutual understanding in exchanging a glance with a gorilla than any other animal I know', he said quietly to the camera. It was a bitter-sweet moment: as the two fellow creatures – man and gorilla – made a playful connection, Attenborough couldn't help but recognise that he might be seeing some of the last of their kind.

The gorillas in that area were on the brink of extinction, their forest habitat being rapidly converted to agricultural fields. But this particular story ended well. The Rwandan government, conservationists and local communities worked together to preserve the gorillas' habitat and Poppy grew up to have many offspring. 'It just shows what we can achieve when we put our minds to it', Attenborough said.[1]

Attenborough's adventures had a deep effect on me as a boy, and this celebrated naturalist has been a constant source of inspiration and wonder ever since. Described as the most 'trusted man on Earth',[2] his calm yet compelling delivery has brought nature into the living rooms of several generations.

Now well into his nineties, far from slowing down, in recent years, Sir David has done some of his most important work. His documentary film and accompanying book, *A Life on Our Planet*, is not just a series of recollections from an extraordinary life but a heartfelt rallying cry against the climate and nature emergency that threatens the planet. 'The natural world is fading', he writes. 'The evidence is all around. It has happened during my lifetime. I have seen it with my own eyes. It will lead to our destruction.'[3]

Attenborough's words led me to reflect on what has changed during my own lifetime, and the rise of industrial agriculture, which has been so integral to the threat that is now facing the natural world.

1965

I was born in 1965, a time of optimism. The world had recovered from the Second World War, technology was making our lives much easier, and the Space Race was in full swing. NASA's Mariner 4 spacecraft had flown past Mars, giving us a glimpse of another planet for the first time. As Generation X, we were born into a world of growth, with the freedom to party. Society was prospering and we felt unstoppable, though there were also growing tensions. The Vietnam War was escalating, and fears were rising over the threat posed by nuclear weapons.

I grew up in the English market town of Leighton Buzzard in Bedfordshire and, from an early age, was fascinated by the countryside and the animals within it. My parents shared a sense of social duty. My father ran a day centre for senior citizens and people with disabilities. He was high-ranking in St John's Ambulance, sat as JP at the local magistrates' court and was legendary for his organisation of youth clubs. But his greatest passion was the Church; in later life, he became Reverend Peter Lymbery, a title that he saw as his crowning glory. Mum was his greatest support.

From writing his speeches and taking minutes to cleaning up after events, Mum was the one who, behind the scenes, did the hard work. She was also a nature lover, tending to the badgers behind the day centre and teaching us children to have respect for nature.

Mum and Grandad would avidly feed the birds in our garden. Among my earliest memories is a stream of starlings descending on the garden to gorge on breadcrumbs. I remember one particularly cold winter seeing a particularly beautiful bird in our garden – a goldfinch – which I identified using *The Observer's Book of Birds* Mum had given me to keep me occupied during a bout of chickenpox. From then on, I was hooked. I recall seeing gulls and imagining they were puffins. I watched crows above the local council building and dreamt they were eagles. I became fascinated by nature documentaries and was soon a keen admirer of David Attenborough.

Mum also taught me compassion towards bees and ladybirds; after all, she'd say, 'they do you no harm'. There seemed to be loads of them back then, and all sorts of other insects too. Whenever we went for a drive in Dad's Hillman Super Minx, the windscreen would be covered in bugs. The fact that the same thing doesn't happen today is a telling sign of how insects have declined.

Back then, I'd spend endless hours roaming woods, riverbanks and canals in search of wildlife, and there were always plenty of farm animals out in the fields. My formative years equipped me with the knowledge to work as a wildlife tour leader; I spent a decade taking nature enthusiasts to some of the most beautiful places on the planet. I never tired of seeing the thrill on their faces when they saw a new type of bird or mammal for the first time. Those years also prepared me for a career in animal protection. By 1990, I was working for Compassion in World Farming, spreading the word about the perils of factory farming. When I learned how husbandry practices that inflicted cruelty on farm animals were also linked to the demise of wildlife, it hit me like a brick.

In 1965, when I was born, the world seemed very different; it had a population of just 3.3 billion people.[4] Climate change wasn't really talked about even though carbon in the atmosphere – correlated with rising temperatures – had already exceeded 320 parts per million, more than in the previous 2.5 million years.[5] Factory farming had been exposed for the first time the previous year, with the publication of Ruth Harrison's explosive book, *Animal Machines*. Compassion in World Farming was yet to be formed; its founder, Peter Roberts, was still farming pasture-raised dairy cows.

At this point, the number of farmed animals produced each year globally for food stood at 10 billion, with meat being eaten much less often than it is now – when my mum started buying chicken for Sunday lunch in the early 1970s, it was seen as an aspirational treat. Although things were changing fast, most meat came from animals that were kept more naturally. Meat was also nutritionally different; compared with a chicken in the 1970s, today's factory-farmed birds have nearly three times as much fat and a third less protein.[6] Back then, they lived longer and more naturally; now they are crammed into huge sheds and given little room to move. Fed diets of concentrated cereals and forced to grow as quickly as possible, they become grotesque parodies of the creatures they should be.

2020

By 2020, the world's population had reached 7.8 billion and carbon in the atmosphere measured 415 parts per million.[7] The world was on track for a temperature rise of around three degrees Celsius by 2100, or higher still if governments failed to keep the commitments of the Paris Climate Agreement.[8]

Our taste for meat had grown to the point where 80 billion land animals were being reared and slaughtered for food each year, a number that was only increasing. For every person on the planet, ten farmed animals were produced a year: a ratio that rose with the trend towards greater meat-eating. Factory farming was the engine room behind our appetite for cheap meat, with two-thirds of the world's farmed animals enduring lives of confinement.

Seeing the reality of factory farming for myself made me weep. These days, I cannot think about factory farms without being reminded of what might happen if artificial intelligence were to get out of control, if robots were to take over the world. For a glimpse of how intelligent, all-powerful machines might treat *us*, we should look no further than today's factory farms.

By 2020, the world's wildlands were in retreat. They had once covered 60 per cent of the planet, but now occupied little more than a third.[9] In the same half-century, numbers of mammals, birds, fish, amphibians and reptiles had fallen by an average of more than two-thirds,[10] and this showed no sign of slowing. At sea, overfishing – including of the vast numbers of fish needed to feed factory farms – had wreaked havoc on marine life. On land, the main causes of decline were intensive agriculture, deforestation and the conversion of wild spaces into yet more farmland.[11]

In an evolutionary heartbeat, what was once wild has become domestic. Half the fertile land on the planet is now farmed, with humans and the animals we rear for food accounting by mass for 96 per cent of all mammals on Earth. Everything else, from elephants and wild bison to badgers and mice make up just 4 per cent. In the avian world, domestic poultry account for 70 per cent of birds by mass.[12] As David Attenborough puts it, 'This is now our planet, run by humankind for humankind. There is little left for the rest of the living world.'[13]

The change in how we eat meat – from occasional treat to everyday meal – has quite literally changed nature. What was for millennia a finely balanced life-support system is now on the brink of collapse, and the biggest single reason for the disappearance of nature has been the great population explosion of livestock.

TOWARDS 2050

To get a sense of how farm animals might define our future, we should focus on what a billion people means. Imagine opening your morning newspaper and reading the headline 'Government to build a hundred new cities the size of London'.

A billion people would be equivalent to a hundred new Londons plus thirty more cities the size of Los Angeles. Having gone from 3 to 8 billion people within half a century, we are set to add another 2 billion to the planet by 2050. The figures are staggering. With more than half the world's people now living in cities, the pressure of urbanisation is often cited as a big environmental problem.[14] Yet the much bigger problem lies not in the cities, but in the vast expanse of countryside used to feed them.

Cities are like a tiny yolk of a sprawling fried egg, with the vast area of land required to grow food for them representing the egg-white. In Britain, about a tenth of land is urban,[15] while 70 per cent is devoted to agriculture.[16] The situation in the EU is similar; urban areas account for less than 7 per cent of land use, while agriculture covers about 40 per cent.[17] In the US, cities cover a mere 3 per cent of land, while its agriculture occupies 44 per cent.[18] And on top of that is the food and feed imported from other continents. For example, despite having one of the world's biggest agricultural land footprints, Britain is still less than two-thirds self-sufficient in food.[19]

The pressure on land has been exacerbated by the population explosion in terrestrial farmed animals. As it stands, a billion extra people mean 10 billion additional farmed animals every year, together with the associated impact on land, water and soil. Furthermore, that ratio – a comparison of UN human population statistics set against Food and Agriculture Organization database (FAOSTAT) figures on slaughtered livestock – looks likely to increase. Meat consumption forecasts[20] charted against projected population increases suggest it could reach more than 13 animals per person by 2050.

As the farmed-animal population rises, so too does the need for cultivated land – particularly when the majority of those animals are raised on factory farms that rely on arable crops for feed. By 2050, global farmland looks set to expand by more than a quarter, which will have a devastating impact on remaining wildlands.[21] Billions more farmed animals will need many more millions of hectares for animal feed. Maize alone could require new farmland equivalent to nearly twice the size of Germany. At the same time, drought and desertification could render a quarter of the world's cropland completely useless.[22] And, as if a decline in arable land weren't enough, global warming could lower crop yields, frustrating attempts to limit agricultural expansion and forcing up food prices.[23]

As the pressure of greater numbers of farmed animals and people increases, some see further intensification as a way to get more from less land. Yet recent decades have exposed 'sustainable' intensification as an oxymoron. Trying to get more from the same amount of farmland erodes its resources and seriously damages the prospects for food production in the future, resulting in what the UN FAO has described as a 'vicious downward spiral'.[24] As David Attenborough puts it, 'The truth is that we can't hope to end biodiversity loss and operate sustainably on Earth until we cease the expansion of our industrial farmland.'[25]

As the footprint of humanity grows, agricultural encroachment and the industrialisation of farming causes irreversible damage to biodiversity, forests, soil and water. More wildlife extinctions follow. Nature is overwhelmed. And as she retreats, nature stops providing essential services like pollination, soil replenishment and carbon sequestration.

EXTINCTION

Throughout human history, *Homo sapiens* has outdone all-comers. From the magnificent bison that once roamed the American plains to the passenger pigeons that once flocked like great rivers in the sky, whatever has stood in our way has been seen off. And with scientists suggesting that we have moved into our own geological era, the Anthropocene, we are now the major force shaping the planet. And we are starting to recognise that one of the biggest victims of the Anthropocene could be ... us.

The last 500 million years have seen five episodes of sudden dramatic loss of biodiversity, and we are racing towards a sixth. Dinosaurs developed after one of the biggest mass-extinction events some 250 million years ago, before disappearing about 66 million years ago. Although the exact causes of previous mass extinctions remain a mystery, volcanic eruptions and asteroid strikes are two prime suspects. The resulting dust clouds probably blocked out sunlight for months if not years, causing plants and plant-eating creatures to die. Heat-trapping gases would have triggered runaway global warming.

The Earth has bounced back from such mass extinctions; though the last recovery took 30 million years,[26] the planet has healed to form the rich biodiversity we enjoy today. And if life is to thrive on our planet, immense biodiversity is essential. Billions of different life forms, each with their own ecological niche, make up the web of life that we depend upon; start removing pieces and the whole is compromised. The interconnectivity of countless plant and animal species makes for what David Attenborough describes as a situation where 'the greater the biodiversity, the more secure will be all life on Earth, including ourselves'.[27]

It was this profusion of life that cradled our own species as we emerged from the forests of Africa 200,000 years ago. We were hunter-gatherers entwined within nature, a fully functioning part of the same ecosystem that sustained us. Then, 10 millennia ago, we settled as agriculturalists and our future became bound up with nature as we worked to grow future harvests. We prospered like never before and had time to devote to culture and society. But gradually, instead of seeing ourselves as part of nature, we came to feel constrained by it; as we have risen inexorably, the natural world has gone into retreat. Our political leaders have reflected our human-centric view of the world, focusing on economics, technology and business rather than an understanding of the natural world. Not surprisingly, they are ill-equipped to recognise the decline of our life-support system, much less to know what to do about it. Yet as we face a greater ecological emergency than at any other time in human history, leadership is desperately needed.

In one human lifetime, the planet has gone from being a bountiful Garden of Eden to a world in decline. At the centre of that decline is the industrialisation of agriculture and our overconsumption of meat. Scientists warn that we have just a few years left to solve climate change, yet production of industrialised meat is increasing, with animal farming responsible for nearly a sixth of greenhouse gas emissions. The use of chemical sprays has led to a decline in pollinating insects. Antibiotics, a large majority of which are fed to factory farm animals, could soon stop working. And then there is the UN's warning that soils could be depleted within sixty years. It is not the precision of predictions that is most important, but the indisputable trend in decline.

In a warming world with more people, more farmed animals and our ecological life-support systems failing, business as usual is not an option. Climate change threatens to throw our planet into chaos. Whole forests could disappear. The Amazon rainforest could turn to savannah or even desert. The world could be lashed more frequently by severe storms, drought, floods and crop failures.[28] Breadbasket regions that we rely on for crops could fail simultaneously, causing massive disruption to food supplies.[29] Low-lying cities and regions could be submerged by sea-level rise, including hundreds in America.[30] Bangladesh faces the threat of disappearance.

In such circumstances, millions of 'climate migrants' would be forced from their homes by extreme weather, crop failure or conflicts over increasingly scarce resources. And as the world goes into a tailspin, it is hard not to imagine former hunter-gatherers turned agriculturalists turning to nuclear weapons. A day that could precipitate a nuclear winter, the like not seen since that meteor strike wiped out the dinosaurs.

We currently stand on a precipice, teetering on the edge; what we do now is of the utmost importance.

'I fear for those who will bear witness to the next ninety years, if we continue living as we are doing at present', David Attenborough has written. With the natural world on course to collapse, the environmental services it brings could falter or fail. If this were to happen, 'it would irreversibly reduce the quality of life of everyone who lives through it, and of the generations to follow.'[31]

RESTORATION

With what we eat affecting the planet so fundamentally, there has never been a greater reason for transforming our food system.

This means fixing both sides of the problem: production and consumption. The way we produce food should move away from industrial agriculture to regenerative, agroecological methods that restore soils and biodiversity. Our consumption would do well to shift from diets that are heavy in livestock products. Nature should be an inherent part of our farms, and where possible it should be encouraged through rewilding, not least of the soil. Great store should be placed on positive, life-affirming ethical solutions to relieve the pressure on a world with limited resources. And alleviating poverty and empowering women and girls will both address population pressure and help build an ethical and decent society. Together with respecting animals and the natural world, such things provide hope for a better future.

Although time is running out, it is not too late for us to leave behind a healthy planet for future generations. Critical to this is appreciating the interconnectedness of the challenges facing humanity. Like David Attenborough, I see a multiplicity of solutions combining to create a genuinely sustainable future. In his words, 'Regenerative farming is an inexpensive approach able to revive the exhausted soils of most fields.'[32] On consumption, he sees a future where 'we will have to change to a diet that is largely plant-based, with much less meat'.[33] Attenborough himself has gradually stopped eating meat. 'I can't pretend it was overly purposeful nor even that I feel virtuous for having done so, but I have been surprised to realise that I don't miss it.'[34] He also anticipates that cultured meat produced from stem cells will play an important role, in addition to proteins made from microbes through fermentation and crops grown in vertical farms. We will explore these methods of food production in the next chapters.

One thing is certain: this isn't a question of people versus animals versus planet. We are all in this together. As Attenborough says, 'It's not about saving the planet – it's about saving ourselves.'[35]

PART FOUR

SPRING

First light on Easter Sunday and the dawn chorus was in full swing. Around our hamlet, avian residents were busy staking their claim to nesting territory. A male tawny owl hooted at the end of a night's hunting. Blackbirds were flushed with new-day vigour, while wrens sang with more power than birds much bigger than themselves. The silvery trill of a robin sprinkled sonic glitter on the morning. As the light came up, a squirrel scurried along the telegraph wire towards me, before thinking better of it and retracing her steps.

The clear dawn sky was almost white but for a band of clouds painting furrows in the sky. The aerial art flared red, then dissolved. Sun broke the horizon. Daffodils had faded. Snowdrops had become a memory. Today was time for the big show: bursts of blossom on the cherry trees, while bluebells carpeted every copse.

The River Rother was calm as it flowed through the limestone arches of the bridge where Duke and I watched mandarin ducks, grey wagtails and kingfishers. Naked oaks along the riverbank were budding with promise, as morning mist hung delicately over riverside pastures. For the grass, the warmth was a spur to start growing, slowly at first. For now, the forty cows behind our house were in the barn where they had spent the winter.

Give it a month and the pasture would be grazed again by our neighbour's cows, and the four hens that had arrived at the house the previous day would be exploring our garden. Until then, our newly *adopted flock were being kept in a shiny coop while they got accustomed to their new surroundings.*

Give it a month and the bare ploughed fields behind our hamlet would be tinged green with the shoots of this year's harvest.

Give it a month and avian migrants would be back from Africa; although as I was to discover, for some, spring had come early. As the sun warmed the day, mason bees busily searched for holes and crevices, dancing over cracks before disappearing for a moment with a flash of hairy abdomen. The first house martins arrived back from Africa, their bright white rumps standing out as they twisted and turned in the air, while swallows swung low over tufty grasslands.

Suddenly, the peace of the day was disturbed by the sound of discord, as a pair of nesting buzzards wheeled in the skies above our hamlet in noisy pursuit of an intruder. The interloper was a large hawk similar to the buzzards but paler and with longer, thinner wings. The stranger's head and body were white but for a dark Lone Ranger mask: I was watching a migrating osprey. This one was back from sub-Saharan wintering grounds in search of somewhere to nest. When I was a budding naturalist back in the 1970s, ospreys were the poster child for successful conservation, a bird driven to extinction yet brought back, first to Scotland then further afield.

I could scarcely believe it: here, for the first time above my home, was a bird long associated in my mind with landscapes restored to life. I was elated. That moment instilled in me an excitement, a renewed optimism at the prospect of a spring rebirth, where the world around me might be restored to its former glory.

15

REGENERATION

RESTORING DIRT TO SOIL

North Dakota, where a herd of cattle is scattered over sweeping prairies with tall grasses covering flat plains and low rippling hills. The cattle are either all black or all hazel. No in-betweens. They move as a group, devouring flourishing grasses and flowering plants as they go. Some break ranks to drink from a still pool, their mouths sending shivers over the surface. Look closely and you'll see wire fencing – not enough to disturb the scene, but enough to keep the cattle moving in an orderly fashion. A gate on an automatic timer soon swings open and the herd moves on to pastures new.

It was talk of regenerating soils that caught my interest in this place, but soil was the one thing that you struggle to see here; as I learned from the man who runs this extraordinary setting, the best way to look after soil is to keep it covered up. And in contradiction of the hymn that we'd sung at my church school – 'we plough the fields and scatter the good seed on the land' – the last thing you'll find on this ranch is a plough. To see the soil here, you have to dig into it and bring to the surface a hidden world, bound by complex root systems that connect this living landscape and hold it fast. Breaking open a clod would reveal worms wriggling in protest – and plenty of them.

Gabe Brown is making waves around the world for how he's doing things; Brown's Ranch, just east of Bismarck, is leading the way in a new wave of regenerative farming. In a world of more people and shrinking resources, an approach that preserves the essential elements of life – soil, water and sunlight – has a lot going for it.

The remote feel of Brown's 6,000-acre ranch belies its proximity to the bustling city of Bismarck. The city was founded in 1872 when this land was the domain of Native Americans and bison; the Northern Pacific Railway was built shortly afterwards to bring in settlers and those drawn by the gold rush in the nearby Black Hills. For thousands of years, successive cultures of indigenous peoples had lived in harmony with the environment and the bison that populated it. With the advent of European settlers, the bison were quickly wiped out by hunting and disease spread by cattle. Today, small herds remain at Theodore Roosevelt National Park and Cross Ranch Nature Preserve.

Brown is trying to see things from the soil's point of view and use his ranch to show how agriculture must reconnect with nature if farming is to survive. He knows all about the hostile world above the soil, one of wind, rain and of being exposed to the elements. But it is what goes on underground that fascinates him, and the key to looking after the soil lies in what is done above it.

Situated in the middle of the Great Plains, the seasons here are variable. Winters are cold and snowy, while summers are hot and humid. Spring and summer can bring huge thunderstorms. In this sparsely populated state, threats of drought, floods, tornadoes and blizzards are part of the landscape, to the point where many farmers are dependent on government aid.[1] Perhaps the variability of the weather has been one of the reasons for Brown's keenness to find ways to protect the soil. His approach to addressing difficulties is to ask, 'Are we solving a problem or just treating a symptom?'[2]

Brown bought the ranch from his wife Shelley's parents in 1991 and now runs it with their son Paul. When they took it on, it was clear that the soil was severely depleted. Organic matter had dropped from an estimated 7 per cent before European settlement to less than 2 per cent.[3] At this point, the Browns were 'conventional' farmers, growing grain and rearing cattle with heavy ploughing and plenty of artificial fertilisers and pesticides. A school class in vocational agriculture led Gabe to join Future Farmers of America, which meant learning about synthetic fertilisers, pesticides, artificial insemination, feedlot rearing and everything else related to industrial agriculture. Like so many farmers, he did what he knew. However, soon after he took over the land, the farm suffered four successive years of weather-related crop failures. The North Dakota climate had hit them hard, leaving them financially desperate, but Brown kept the faith and broadened his knowledge.

Desperate circumstances and a thirst for knowledge encouraged Brown to do things differently. In his book *Dirt to Soil* he recounts a light-bulb moment when the Canadian rancher Don Campbell said, 'If you want to make small changes, change the way you *do* things; if you want to make major changes, change the way you *see* things.'[4] Until then, he had been making minor adjustments to the ranch while praying for big results. If he was to dig himself out of his hole and stay in business, he needed to look at things entirely differently. Those four years of crop failure turned out to be the best thing that could have happened to him. 'They forced us to think outside the box, to not be afraid of failure and to work with nature instead of against her. They sent me on this journey of regenerative agriculture', he wrote.[5]

The first sign of Brown's success in regenerating his farmland ecosystem was the appearance of earthworms. 'I used to joke that we could never go fishing because there weren't any earthworms'.[6] During the years of crop failure, he'd left the soil covered and reduced the amount of artificial fertilisers and pesticides because money was so tight. The results were transformative. When he sank his shovel into the soil, there were worms and the earth was darker and richer, with better structure – it looked like chocolate cake. It held more organic matter and water – it was alive. 'Thanks to the crop failures, I changed the way I looked at our land. Unfortunately, the good Lord had to slap me four times before I woke up!' he said.

Since then, Brown hasn't looked back. Soil health and diversity have become his priorities. His land hasn't been ploughed since 1993. Ploughing 'destroys soil structure, the home for soil biology', he said.[7] Here, diversity of crops and animals is the name of the game. Synthetic fertilisers and general pesticides are no longer used, and although he does admit to using some herbicide, he is working hard to eliminate it. And Brown's Ranch is GMO- and glyphosate-free.

Brown stepped off the intensive treadmill, but I was keen to know why he thought other farmers stay on it. 'Farmers and ranchers get exposed every day to fertiliser dealers and chemical dealers and implement dealers and government farm policy', he said. 'They don't get exposed to biodiversity and regeneration and how soil biology really functions to supply nutrients.'

Brown reckons a healthy grassland range has over one hundred species of grasses, herbs and shrubs, which is why, in contrast to intensive agriculture, his croplands host a diverse range of plants and crops. This diversity provides the food that feeds soil life, which in turn supplies nutrients to crops. 'Where in nature do you see a monoculture?' he asked.

Another key to healthy soil is keeping the ground covered at all times, using cover crops and crop residue. 'These provide the armour to protect the soil from wind and water erosion', he told me. Soils are not left exposed to the elements between main crops, with cover crops maintaining the nutrient exchange between roots and soil microbes, keeping the soil's food web alive. They also help to integrate crops with farm animals, with the various cover crops being grazed by cattle during the winter. Keeping cattle outdoors in this way helps keep them healthier. They deposit dung as they go, adding to the soil and supplying nutrients to subsequent crops. And it makes things easier for Brown and his family. 'We don't have to haul manure out of corrals and onto the fields,' he said. 'It's a win-win-win situation.'[8]

Some 350 pairs of cows and calves roam this ranch, along with up to 600 yearling cattle. The calves graze with their mothers during the winter, learning which plants to eat and how to use snow for drinking water. In the spring, they are weaned from their mothers using a fence that prevents them from nursing but still allows them to see each other and to touch noses. Much more welfare-friendly, it avoids the separation anxiety seen in forced weaning, when calves are removed from their mothers and taken elsewhere.[9]

The cattle are integrated with crops and Katahdin hair sheep, with poultry being part of the mix since 2011. Tamworth and Berkshire pigs have also been added to the farm, and the young pigs get up to half their food from perennial pastures. Their rooting around for food encourages grasses, herbs and other plants to germinate, which creates a healthier ecosystem.[10] Brown's cattle and sheep are now grain-free, being fed entirely on grass and other forage – just as nature intended. Reflecting on his past practices, he asked himself why he was feeding animals grain, when that's not how they evolved. So he stopped doing it and now reaps the benefit. 'When these animals are fed a forage-only diet like they are designed, they provide a much healthier end product for the consumer.'[11]

Cattle and sheep are integral to a thriving mix of species on the farm, grazing a 'diverse sward' – farmer-speak for a diet of pasture that contains a variety of grasses and herbs. And as they move across the landscape, they help to regenerate soils.

Brown sees industrial farming's removal of farmed animals from the landscape as a 'tragic flaw'.[12] In the US, you can drive for hundreds of miles without seeing a single animal out in the fields. Instead, they are locked in warehouses and fed on grain that could otherwise be feeding people. It's a far cry from the natural ecosystem of the Great Plains.

'Nature does not function without animals', is how Brown sums things up in *Dirt to Soil*. He explains how grazing animals have a symbiotic relationship with their environment. As they feed, the plants release carbon through their roots, attracting soil microbes, which in turn supply nutrients for new growth.[13] Studies have proven the beneficial role of rotational grazing on soil regeneration, including carbon storage.[14] As Professor David Montgomery puts it: 'Reintegrating animal husbandry onto farms offers a powerful tool for regenerative agriculture.'[15]

Grazing farm animals can turbo-charge soil regeneration. Add to this the myriad of insects, birds and other wildlife, and you have what Brown describes as a 'very healthy, optimally functioning ecosystem'.[16] He sees grazing animals as being the 'accelerator' for rebuilding soil health – 'they will jump it to another level. It's that biting of the living plant that causes the plant to slough off root exudates, in order to attract more microbiology to regrow.' That and the hoof action of the animals, which tramples organic material into the soil along with their dung and urine. 'It's that exchange of biology that stimulates the nutrient cycle.'

Brown recalls how bison on these plains were once followed by a diversity of other animals, including birds. Today, he tries to emulate that natural relationship on his farm with chickens. His 1,500 laying hens follow the cattle on the pasture about three days later, roaming free during the day and roosting and laying eggs in portable 'eggmobiles'. They spend their days pecking at cowpat-peppered pastures, picking out bugs and turning up worms and seeds. Dashing for spiders and flies. Their natural food is supplemented by residue from the farm's grain crops, along with seeds from weeds that may have been harvested with the grain.[17]

Unlike cattle and sheep, chickens are naturally attuned to eating grain, but the amount they consume is tempered by what they find on the range. By using those grains from the farm that would otherwise go to waste, Brown uses the leftovers from one enterprise – his crops – to fuel another – his chickens. The hens' longevity is testimony to the life they lead; an industrial layer would seldom reach more than eighteen months of age before going for slaughter, while some of Brown's hens live for seven years.[18]

Brown also has a thousand chickens of a different breed for meat, and these are also out ranging, moved daily in portable pens. Brown says they get 'most of their diet from what they find in the pasture', topped up with a bit of grain. He believes in allowing a cow to be a cow and a chicken to be a chicken – it's important to him that animals on his farm can do what comes naturally to them. 'Whether it's the cattle moving across the prairie in a herd or a chicken chasing after that tasty grasshopper, our animals are grown and enriched in a stress-free environment.'[19]

Brown doesn't use growth hormones and sees antibiotics as unnecessary because the animals' well-being is taken care of by the way they are kept. 'When livestock and man coexist in a stress-free environment, it provides for a higher quality of life for all. And that shows in our produce.'[20]

Looking after the soil means that Brown is better placed to look after water resources and maximise the use of sunlight on his farm. From less than 2 per cent organic matter in 1991, the soil now contains up to 7.9 per cent, which he puts down to the diversity of crops, animals and microbes working together.[21] An acre of his soil held about 40,000 gallons of water when he took over, but regeneration of organic matter has boosted this to 100,000 gallons.

Restoring farm animals to the land has played a big part in this – acting as the 'accelerator' – as has his use of cover crops.[22] As for the future? It's about 'regenerating landscapes for a sustainable future', he said.[23] 'Not only ours, but future generations as well.'

LIKE PIGS IN CLOVER

'Now somewhere in here, there are some pigs', my host said, flinging open the gate to what looked like an empty field of grass. We bumped along in his Land Rover and soon came to fourteen Tamworth pigs lying on a bank, knee-deep in clover. They had been feeding all morning and were now resting during the midday heat, lying still but for the odd flick of an ear or tail. They had no rings through their noses, a cruel device that would snag their sensitive snouts and prevent them from rooting up the grass. They had short ginger coats with neatly pointed ears, without tags. These pigs were comfortable in their surroundings and doing what comes naturally to them: eating.

I was about to learn what pigs really like to eat and how wrong it is to use the phrase 'like a pig in muck'. Most modern-day pigs live in muck, the vast majority on factory farms, on bare concrete or unyielding slats, which is why we assume that pigs want to be in muck. Furthermore, it is often all we allow them to do. Yes, they like a wallow, just as we like a bath, but what they really want is to be free to eat as much clover – their favourite food – as they like. Hence the true English idiom 'like a pig in clover', a description of being truly contented.

Losing sight of a pig's real nature has given rise to one of the trickiest problems in the move away from industrial agriculture – how to avoid them being grain-guzzlers. Most commercially reared pigs are kept intensively; most of the rest live outdoors in porcine monocultures, which, while better for their welfare, is not quite the integrated solution of a regenerative farm. Either way, they are often fed vast quantities of cereals, soya and fishmeal. Left to their own devices, they would root around in small groups, building nests for their piglets at the edges of woodland while eating just about anything. Pigs are nature's recyclers and have long been domesticated for their ability to live on table scraps. How then to best integrate them into a regenerative farm was a question answered on a chance visit to a farm in Wiltshire: pasture pigs.

At Horton House Farm in the Vale of Pewsey, between Salisbury Plain and the Marlborough Downs, Jonny Rider and his family raise a hundred pigs on grasses, clovers, herbs and wild flowers. Rider's family have been farming here since 1954, and he took over the farm from his father in 1999, soon after returning from Wye Agricultural College. He now farms with his wife Rachael, and their four children are all keen to come back to the farm to work.

Quiet and informal, with close-cropped curly hair, forty-nine-year-old Rider showed me round his 400-hectare farm, where cattle, sheep, chickens, goats and pigs move through the blocks of pasture in rotation. A self-confessed Tamworth pig enthusiast, the way he keeps his pigs is cutting edge – nature, soil and welfare-friendly all at the same time. Rider lets his pigs suckle with their mother for five to six months – six times longer than their intensively reared cousins, who are separated from the sow within a month. They get to live twice as long, too; intensive pigs are pushed to grow unnaturally fast and are slaughtered when they are barely six months old.

During the grazing months, the pigs at Horton House get their nutrition from pasture. I learned from Rider how the pigs root up any thistles before settling into any new pasture and feasting on the variety of clover, grasses and herbaceous flowers. When the grass stops growing, they are moved to winter quarters with the cows, where the piglets playfully bury themselves in the straw beside the cows and even eat cowpats. During the winter, the pigs' diet is supplemented by waste milk from Rider's pasture-fed dairy, where no grain or other concentrates are used to boost milk yield. As well as being fed milk that would otherwise be thrown away, the pigs get through the winter with sprouted oats and lucerne grown on the farm.

Rider estimates that his pigs will get three-quarters of their entire diet from pasture in their lifetime. The economics of longer-living grass-fed pigs adds up because pasture is much cheaper than commercial cereal and soya-based feed. 'Grazed grass is a tenth of the price of organic feed', he explained, so he doesn't have to make the pigs grow ultra-fast to turn a living.

The pigs can even be found grazing in the same fields as cows, where they help to cleanse the pasture of potential parasites, keeping the farm healthy. 'They sanitise the pasture … and distribute the dung around', Rider said, which helps with natural soil fertility. All of this is a far cry from industrial farms, where most pigs never see the light of day and those that do are often overstocked on bare soil.

As pig farms go, Rider's is small, producing a modest hundred pigs a year. However, most pig farms don't also have cattle, sheep, goats and chickens. Rather than a single monoculture, he produces a range of products in harmony with nature. And he's not alone; a growing number of farmers are now keeping pigs on pasture. Somerset farmer Fred Price ran a conventional farm before turning to regenerative agriculture and keeping thirty Tamworth breeding pigs, with as much as four-fifths of their feed coming from pasture. And Simon Cutter's regenerative cattle farm at Ross-on-Wye also keeps pigs on pasture and food waste.[24]

Rider's wife Rachael explained why they farm like this: It's 'because we want to farm for the children and have a safe, nice environment for them to grow up. We also want to feed them really healthy food.'[25] My trip to the Vale of Pewsey had taught me that the secret of keeping pigs regeneratively is, quite literally, to keep them like a pig in clover.

UNDERGROUND LIVESTOCK

'Regenerative agriculture' has become a buzzword term for practices that increase soil fertility and carbon sequestration, biodiversity and water conservation, but does it always require farm animals? There are plenty of examples of degraded soils being revitalised by the addition of animals to the land, which improves the welfare of the animals who would otherwise be confined in factory farms while also restoring soil health. Yet what if farming decided to dispense with farmed animals altogether? Could crop farming still be regenerative?

In seeking an answer to this question, my research drew me to Ohio, where David and Kendra Brandt farm regeneratively on a 1,200-acre farm near Carroll.[26] I had hoped to ask them if I could come and visit personally, but Covid made that impossible. The reason for my keenness was that they have found a soil-friendly way of producing common commodity crops – corn, soya and winter wheat – that are usually produced using ploughing, artificial fertilisers and chemical pesticides. But their land hasn't seen a plough since 1971.

By ditching the plough, keeping things rotating around the farm and keeping the soil covered with cover crops, David Brandt has all but dispensed with synthetic fertiliser. According to fellow regenerative agriculture advocate Gabe Brown, his soils are amazing. 'You can stick a spade into any of David's fields and reveal eighteen-plus inches of dark chocolate cake-like topsoil,' Brown wrote. 'If you step over onto neighbouring property that has been farmed conventionally with tillage and you stick in your spade, you will find tight, yellow clay. The contrast is stark.'[27]

During spring and summer, Brandt's fields look the same as those of his industrial neighbours, but in autumn, when adjacent fields are bare, his are full of swaying cover crops. During the off season, he grows plants like sunflowers and radishes, mixed with various grasses that form a layer of vegetation over his soil, just as nature intended. 'What we're trying to do is do like Mother Nature does', he told *Farm and Dairy* magazine.[28]

Brandt started planting cover crops to control soil erosion in 1978. After all, the worst thing for soil is exposure to wind and rain, and cover crops keep the soil covered up. According to one report, he uses ten different varieties of plant in his cover crop communities, a 'cocktail' aimed at having plants of different heights and root patterns, both nitrogen-fixing legumes and non-legumes. However, the practice is rare in the States, used on only 1 per cent of American farmland.[29]

Brandt didn't start out in regenerative farming, and his no-till experiment developed more out of necessity than choice. He grew up helping on the family farm, milking cows, rearing pigs and growing potatoes. He married Kendra in 1966, before being posted to Vietnam as a sergeant in the marines for two years. While he was away, his father was killed in a tractor accident. The family were forced to sell the farm, but Brandt's ambition remained undaunted. He took on a farm tenancy but couldn't afford a plough so went no-till, and it turned out to be the making of him.[30]

Now seventy-five years old[31] and something of a rock star in regenerative farming circles, Brandt is recognised as a pioneer in soil conservation.[32] According to scientists writing in *International Soil and Water Conservation Research*, studies on his farm 'convincingly demonstrate' that no-till agriculture using rotations and cover crops 'significantly increases carbon sequestration, improves soil health, increases crop yield and enhances ecosystem services'. The paper goes on to conclude that 'Conservation Agriculture, including no-till, crop rotations and soil-plant diversity, is becoming a global movement transforming how food is produced and how to better manage the ecology in agricultural landscapes. It is proving to be the central approach to global sustainable agriculture.'[33]

Unlike most regenerative farms, Brandt's fields aren't roamed by cows, sheep or chickens – but that doesn't stop him jumping from his utility vehicle with a shovel to inspect his 'livestock underground'.[34] He's referring to the worms, grubs and bacteria that are thriving under the surface, drawing dead stuff from crops and depositing it as nutrients and carbon in the soil.

Back in the 1970s, the carbon content of Brandt's fields was less than 0.5 per cent, just like the surrounding farms. In the decades since he began using cover crops and farming regeneratively, this has increased to 8.5 per cent, a clear 2 per cent higher than nearby native non-cultivated soils. He credits the radishes in his cover crop for the improvement. Their deep-penetrating roots increase phosphorus levels and help break up the soil. And when they decompose, they release nitrogen and other nutrients.[35]

Soil carbon doesn't feed plants directly, but studies show a close relationship between soil organic matter and crop yields – increasing the amount of organic soil carbon enables crop yields to be maintained with less artificial fertiliser. One study of wheat, rice and maize production in developing countries concluded that increased soil carbon can boost yields as well as sequester carbon from the atmosphere, helping food security and climate change mitigation.[36]

While he hasn't gone organic, Brandt saves a lot of money by reducing chemical inputs on his farm. Compared to his neighbours who farm conventionally, he uses less than half the amount of herbicide, a fifth of the diesel and a tenth of the artificial fertiliser, while obtaining consistently better yields.[37] He and his ilk demonstrate that by adopting no-till farming and using cover crops and a diversity of rotational crops, regenerative farming can work without livestock. As the leading soil advocate David Montgomery puts it, 'ditch the plough, cover up and grow diversity'.[38] With or without farm animals, the need for nature-friendly farming has never been more urgent.

During the American Dust Bowl era, William Albrecht, professor of soils at the University of Missouri, wrote, 'The Nation should be made aware of the rapid rate at which the organic matter in the soil is being exhausted … The maintenance of soil organic matter might well be considered a national responsibility.'[39] Three-quarters of a century later, Montgomery, professor at the University of Washington, wrote, 'We need to see soil carbon as a societal investment account – humanity's planetary nest egg. For if we restore it now, our descendants can reap perpetual dividends. Unfortunately, we're still flushing it away.'[40]

As the combined effects of a heating planet, the collapse of nature and the decline of our soils sound an increasingly urgent alarm, there has never been a better time for us to pay attention to the 'livestock underground'.

GROWING UNDERGROUND

Below the bustling streets around Clapham Common in south London, a new food revolution is taking place. Deep underground, tunnels originally built as air raid shelters during the Second World War are producing the most extraordinary harvest of herbs and salad leaves. Tunnels big enough to harbour 8,000 Londoners now play host to tray upon tray of pea shoots, rocket, red mustard, pink-stem radish, garlic chives, fennel and coriander, all supplied to markets and wholesalers across London.

A hectare of gloomy tunnels has been converted into slick growing spaces using LED lighting with a pink hue, in what is claimed to be the world's first subterranean farm.[41] In this futuristic environment, the rows of trays, reaching from floor to ceiling, are nurturing seeds from sapling to harvest.

The secret behind growing food without soil or sunlight lies in a technique called hydroponics: using water to deliver nutrients to the roots. Deep in these tunnels, peas sit on a recycled carpet, their roots growing through the matting to a shallow water bath. Nutrient-carrying water is pumped through the growing trays. As most of it isn't taken up by the roots, it is captured, filtered and reused; as a result, 70 per cent less water is required than in conventional open-field farming. Sunlight needed for plants to photosynthesise and grow is replaced by the wide-spectrum LED lighting. Just like the Tube train that rumbles four storeys above the growing space, this Clapham farm operates all year round, immune to the seasonality that limits open-air harvests.[42]

Growing Underground, the company behind the venture, has received the backing of celebrity chef Michel Roux Jr, who lives directly above the farm and has joined its board of directors. He's enthusiastic about the prospects for growing greens underground and said on a promotional video, 'It's something that I had to do ... the market for this is huge!'[43] Roux's Michelin-starred restaurant, Le Gavroche in Mayfair, is reported to be on the company's client list.[44] Its 550-square-metre area aims to grow about 20,000 kilograms of greens a year.[45]

Steven Dring and Richard Ballard founded Growing Underground because of a desire to create a business that would help overcome the challenges of feeding people sustainably. They saw it as a way to meet demand for locally grown, sustainable produce. And in growing food in the literal 'heart' of the community, they have taken 'local' to a whole new level. 'We've got enough food to feed the planet, it's just in the wrong place. We started looking into that and how to make it into a business', Dring told the *Independent*.[46]

Rather than being a threat, Dring expects the relationship between Growing Underground and open-air farms to be a collaborative one. With a growing population and a finite amount of land, moving production underground frees up space above ground for bulkier crops.[47] John Shropshire, the man at the helm of one of Britain's biggest salad- and veg-growing companies, G's Fresh, is collaborating with Growing Underground. He sees a lot of sense in supplying urban markets with salads and herbs from underground suppliers. He also agrees with Dring that not everything can be produced underground. 'Wheat and other starchy cereals will likely be too energy-intensive for indoor production. Anything that packs a lot of starch will need a lot of energy.'

THE SKY'S THE LIMIT

Farming looks set to hit new heights thanks to AeroFarms, a US-based firm that among other projects is behind a 90,000-square-foot vertical indoor farm in Abu Dhabi, among other projects. Claimed to be the 'world's largest', its environmental credentials include using 95 per cent less water than open-air field agriculture and being totally pesticide-free.[48]

The company, a leader in vertical farming since 2004, uses a patented technique called 'aeroponic' growing. Seeds are placed on a recycled plastic mat, with the plant beds stacked vertically and bathed in LED light. Water and nutrients are sprayed on as a mist, further reducing the amount of water needed. Described by one journalist as an 'ambitious, almost fantastical, manifestation of agricultural technology', proponents of vertical farming call it the 'third green revolution', likening the developments to those made by Apple and Tesla.[49]

From its base in Newark, New Jersey, the company promises crop yields that are seventy times greater than conventional open-air agriculture.[50]

Aeroponics, hydroponics and, as we'll see in the next chapter, the promise of growing meat from stem cells, all have the potential to take industrial farming out of the countryside. The demise of the natural world has been a direct consequence of the widespread adoption of industrial techniques in farming, which the environmental pioneer Rachel Carson foresaw in her seminal book, *Silent Spring*. Carson was born in a world where industry and the countryside existed side by side, but during her lifetime the lines became blurred. The ways of industry were adopted in the countryside, with devastating consequences. She shone a light on the effects of spraying the countryside with chemicals, part of agriculture's new industrialised approach.

Innovative agricultural technologies like aeroponics and hydroponics are unlikely to make a dent in the global food crisis, at least for now. Countries that face the highest food insecurity simply can't afford expensive new technology. However, these developments point to a future where food can be produced within urban communities and without the environmental consequences of industrial farming. In terms of cutting down food miles and increasing local food security, it can only be a good thing. Along with rooftop and urban farming, it has the potential to boost food supplies for cities in an uncertain future. After all, climate change and subsequent sea-level rise will cause agricultural land to shrink, just when we need it most.

16

RETHINKING PROTEIN

MILK'S MORAL MAZE

I was walking through waist-high grasses and past gnarled trees and springy bushes in the Brazilian Midwest. The air was fresh. A big bee buzzed by. This was Emas National Park and a pristine patch of the Cerrado, the grassland that once covered much of the Brazilian interior. On the other side of the road, the land was barren: a vast flat prairie of uniform farmland, with little else but bare, newly planted fields. There were no hedges, no field boundaries, just one crop: soya.

I was witnessing first-hand how wildlife was disappearing from one of the most biodiverse regions on Earth, and it made me stop and think: was I partly to blame for this devastation, thanks to the soya milk in my tea?

The question goes much deeper than my favourite hot drink. Way back in the 1970s, Compassion in World Farming's founder Peter Roberts was an early pioneer of soya products, importing them for human consumption. His company, Direct Foods, sold a range of plant-based innovations made from soya that included Sosmix and Sizzles, the latter mimicking the smoky taste of bacon.

Peter had been a dairy farmer until he became concerned at the rise of factory farming. By 1967, he'd given up his cows and started Compassion as well as a thriving soya business. Back then, he was at the cutting edge of awareness about milk and the ethical questions surrounding it. He was an advocate of rethinking our attitude to protein, seeing it as more than just foods sourced from animals. As well as his food brand, he also set up a health food shop, The Bran Tub, which to this day sells a range of plant-based milks. Having worked for Compassion since 1990, I am proud to have followed in his footsteps.

While I was writing *Farmageddon*, my book inspired by Peter's vision, I saw a lot of mega-dairies, not least in California's Central Valley. I discovered how it was common practice to confine up to 12,000 cows to a single dusty pen, without a blade of grass in sight. Instead, the cows are fed grain, which can upset their stomachs. A typical intensively farmed cow is bred to produce so much milk that, in peak lactation, her body performs the metabolic work equivalent to a human running a marathon every day. No wonder the cows on these 'farms' look like they're walking in slow motion.

During my time in California, I heard how pollution from mega-dairies gets into groundwater and the air; for some residents, bottled drinking water is standard, while asthma is common.

Keeping cows in grassy fields is undoubtedly better for their welfare, providing fresh air, sunshine and the ability for them to live naturally. Keeping ruminants on grass also helps avoid the serious animal welfare problems seen in super-high-yielding breeds, whose overproducing udders require intensive concentrated feed.

Pasture-based dairying is, without doubt, better for cow welfare than intensive mega-dairies, but it still relies on breaking that most basic bond between mother and offspring. To give milk, a cow has to be made pregnant and have a calf, which is then taken away from her at birth. Yuval Noah Harari, author of the bestselling *Sapiens*, described what this separation can do to mother and calf: 'To the best of our understanding, this is likely to cause a lot of misery, a lot of emotional pain, to both the mother and the offspring.'[1]

Apart from a small number of 'calf-at-foot' dairies where mother and calf spend the first few months together, the dairy industry is built on breaking this most basic bond.

Fuelled by a mix of animal welfare and health concerns, interest in plant-based milks has been rising rapidly. Globally, the plant milk market was valued at more than $16 billion in 2019, up from $7 billion in 2010 and set to reach more than $40 billion by 2025.[2] In the US, the $2.5 billion plant milk market grew by 20 per cent in the year to 2020, with more than a third of households using them.[3] In Britain, almost a quarter of people used plant-based milks in 2019, up almost a fifth on the previous year, with the increase particularly marked in younger consumers.[4] Industry analysts put the rise in popularity of these milks down to a surge in 'flexitarian' or 'reducetarian' consumers, suggesting that eating less dairy and meat is becoming mainstream.[5]

David Sprinkle, research director for the market intelligence company Packaged Facts, commented: 'Vegetarians and vegans together account for less than 15 per cent of all consumers and their numbers do not grow very rapidly, but a growing number of consumers identify themselves as flexitarian or lessitarian, meaning that they've cut back on their consumption of animal-based foods and beverages. It is this group that is most responsible for the significant and ongoing shift from dairy milk to plant-based milk.'[6]

Plant milks are made from a growing list of crops, with soya, almond, oat, rice and coconut among the most popular. While they obviously sidestep animal welfare concerns, I was keen to find out whether the plant milk in my tea was having a detrimental impact on the environment.

As I saw in Brazil, industrial soya production is responsible for the demise of whole ecosystems. While there, I took a flight in a small plane from Sao Felix, a place the locals call 'the end of the Earth'. I flew over endless rainforest and saw the mighty Araguaia River winding like a giant serpent. And as we flew further south, things started to change: the rainforest began to show bare patches, small ones at first, then bigger. Then great chunks were gone, until the rainforest was an island in a vast sea of crops, and then it was gone altogether. I looked down at a never-ending prairie of soya, transfixed, feeling like I'd just watched the lungs of the Earth disappearing.

In Brazil, industrial soya is expanding by hundreds of thousands of hectares every year. Monocultures of crops like soya don't have the natural pest-control defences of mixed farms, so rely on high amounts of pesticide. Brazil accounts for a fifth of global pesticide use,[7] and nearly half of it is sprayed on soya. Chemical pesticide use and associated poisoning is so prevalent there that, on average, every two and a half days, someone dies.[8]

My fears about soya were summed up by the film-maker and academic Raj Patel, who said, 'If you are vegetarian and you walk around with your halo of virtue, but you are eating tofu that comes from Brazilian soy, then you're just as complicit in all of this as if you are eating the beef fed on Brazilian soy.'[9]

When I delved deeper, I discovered that only a fraction of soya is used to feed people; though it contains all the essential amino acids needed for human nutrition, the vast majority goes for animal feed.[10] As Peter Roberts recognised half a century ago, we should be using this 'wonder crop' to feed people rather than factory-farmed animals. However, the European Union still imports about 35 million tonnes of soya a year, nearly half of which comes from Brazil, mainly to feed industrially reared farm animals.[11]

I was keen to discover whether the soya used in plant milks came from the deforested plains of South America. And in trying to answer this question, I learned that provenance – where the ingredients come from and how they are grown – is the biggest single factor when it comes to the ethics behind a brand.

Glancing across the supermarket shelves in Britain, one brand that leaps out is Alpro. It's an interesting example of how the lines between the plant milk and the dairy industries have been blurred – since 2017, Alpro's parent company, WhiteWave, has been owned by the dairy giant and French multinational, Danone.

According to the company's product information, the soya used in Alpro's products is non-GM and doesn't come from deforested areas. Half of it is grown in Europe (France, the Netherlands and Belgium), while the rest is from Canada. Alpro also has a policy of sourcing from farms that practise crop rotation, which is more environmentally friendly than the chemical-driven monocultures

I saw in Brazil.[12] The company's products are distributed under the brand names Alpro, Belsoy and Provamel, the last using organic soya grown in the EU. I also came across another soya milk brand, Sojade ('So Soya!'), which uses soya beans grown organically in France.

While the soya milks that I examined in supermarkets and health food shops don't seem to be implicated in South American deforestation, how do they fare in terms of other environmental considerations? According to 'life cycle analysis' data, which looks at factors from field to fridge, soya milk needs 61 per cent less land than the dairy equivalent, produces 76 per cent less greenhouse gas emissions and is four times less polluting of water.[13] Although far from exhaustive, my simple survey discovered that planet-conscious consumers can choose environmentally friendly soya milk by looking for non-GM, locally sourced and preferably organic products.

Almond is another popular plant-based milk, and I glimpsed the dark side of its production in California's Central Valley, where vast monocultures of crops are interrupted by animal factories or huge fruit and almond plantations. I was standing in an almond grove with my writing partner Isabel Oakeshott. We were surrounded by 60 million trees arranged in perfect rows that stretch the best part of four hundred miles. Wildlife was conspicuous by its absence, but we could hear the low thud of a distant helicopter spraying pesticides to keep nature at bay, part of a daily chemical assault on the landscape by aircraft, weird-looking landcraft and men in protective suits.

We learned that California's crops rely on the mass migration of 40 billion bees a year – on the backs of thousands of trucks, the world's largest man-made pollination event. After six weeks of pollinating the crops, the bees and their hives are loaded back onto the trucks and taken to the next eco-stricken area. Standing in a state that is responsible for producing 80 per cent of the world's almonds, I couldn't help but wonder whether, by drinking almond milk, I was doing the right thing.

Conventional almond production is heavy in pesticides and presents a trade-off between water use and climate; a litre of almond milk needs seventeen times more water than dairy, while greenhouse gas emissions were ten times less.[14] I came across Blue Diamond 'Breeze' almond milk in my local supermarket. Their almonds are grown in orchards by a 3,000-strong grower cooperative in California.[15] None of their almonds are organic and they wouldn't tell me what pesticides they used, but the company did tell me that they 'abide by California regulations', some of which I'd seen first-hand during my time in the Sunshine State.[16] I looked at other almond brands in local supermarkets, Alpro again being among the most visible. The company says its almonds come exclusively from small farms around the Mediterranean.[17] Its organic brand, Provamel, uses only organic almonds from Europe. Another prominent brand, Rude Health's almonds are also organic, mainly from Italy.

So I'd found that, with care, you can find soya and almond milks that are OK for the environment, but what about some of the less well-known plant milks?

One type that burst onto the scene recently and has since made it big is oat milk. It all started when photos of Adam Arnesson, a Swedish farmer who was 'not your usual milk producer', circulated in the media. For a start, he 'doesn't have any dairy cattle'. The *Guardian* picked up on the fact that he had switched from growing cereals for animal feed to producing oats for milk. His switch was supported by the Swedish drinks company Oatly, which planned to work with farmers to demonstrate the environmental benefits of crop production over livestock farming.[18] When compared with dairy, oat milk has around 80 per cent less climate impact and uses half the energy.[19]

So in increasingly environmentally conscious times, will milk from oats, soya or almond become the favourite, or will supermarket shelves be dominated by a complete newcomer?

The San Francisco start-up Perfect Day says its fermented milk, produced through a combination of yeast, cow DNA and plant nutrients, tastes identical to cow's milk. 'There are no test tubes in our fermentation process – it's just like brewing craft beer', said co-founder Ryan Pandya. 'The assumption was that we'd be hated by the dairy industry, but it's actually the opposite. They see us as the long-term solution to problems they didn't even know were solvable.'[20]

Conventional dairy problems include high levels of greenhouse gas emissions, water use and land for grazing or growing crops for feed. A study by academics at the University of the West of England suggested that fermented milk could cut climate impact by up to two-thirds, while reducing water and land requirements by more than 90 per cent.[21]

My research into the ethics involved in dairy, plant milks and futuristic fermented milk reaffirmed the wisdom of Peter Roberts in putting his faith in plant-based alternatives as a more compassionate way of feeding the world. For those who continue to drink mainly milk from cows, choosing organic or pasture-fed milk from regenerative and calf-at-foot farms, and less of it, is the way forward. It goes without saying that plant milks avoid the animal welfare issues associated with most dairy farming, but care is needed to avoid getting plant milks from industrial farming. To be sustainable, plant farming needs to be regenerative, too. So, I'd come through the moral maze of milk with clear advice for ethical consumers: always check that your choice doesn't involve industrial farming with lots of chemicals, particularly from South America. And for peace of mind, choose organic.

WILL GOATS (AND OTHER VEGANS) INHERIT THE EARTH?

Springtime in North Wales and the streets of the seaside town of Llandudno were silent, apart from the sound of hooves on concrete. With the town in lockdown due to Covid-19, a herd of goats had spotted their chance to stage a takeover. They descended from the Great Orme, a limestone headland north-west of Llandudno, into the centre of the Victorian town.

The tourists and townsfolk who would normally have mingled amid cafes, bars and shops were replaced by horned, shaggy Kashmiri goats, eager to taste local hedges, lawns, flowers and other forbidden delights. They had once been a gift from Queen Victoria to Lord Mostyn, who owned the Great Orme. Now considered wild, the local council declared itself powerless to stop them wandering the streets.[22]

During a most difficult time for people, the goats provided a talking point that lifted flagging spirits. Yet their arrival in the town centre wasn't the only big change in the 'Queen of the Welsh resorts'. Just a couple of months before, I'd walked those same streets and had been struck by another dramatic arrival: veganism.

Fish and chips

Walk down almost any British high street in 2020 and you'd find companies falling over themselves to show that they were participating in the new food sensation. Llandudno is far from the 'hipster' epicentres of London, Brighton, New York or Berlin. This is a town of modest population, swollen by tourists whose guidebooks list eating fish and chips as one of the main things to do.[23]

For the quarter of a century I'd been coming to the town to see a band called the Alarm, a vegan meal had always consisted of fish and chips, only without the fish. Now the local chippy had a bespoke vegan menu including a dozen plant-based options, while a handwritten note on the window advertised 'vegan Magnums'. Along the road, takeaway bakery chain Greggs had a window devoted to advertising its new vegan 'steak bakes', while in-store promotions proclaimed that their vegan sausage rolls were 'the nation's favourite'. This all seemed a far cry from the ridicule that was previously heaped on anything that advertised itself as vegan. Quite the reverse. It was *that* vegan sausage roll which boosted Greggs' fortunes in 2019, when boorish TV host Piers Morgan spat one into a bucket, sending sales and the company's share price soaring. Greggs really couldn't have paid for a better publicity stunt! Supermarkets like Marks & Spencer, with its 'Plant Kitchen' range, were also broadening their offering, introducing new plant-based inventions at amazing speed. In the same store, the racks of lifestyle magazines included a selection of prominently displayed vegan monthlies. And that evening, I turned on my hotel television to see a Birdseye advert for meatier than meat plant-based burgers.

The Eagle has landed

It was a major coup for veganism when KFC – globally dominant with its fried factory-farmed chicken – devoted an entire window of their restaurants to a giant poster that proclaimed, '11 herbs and spices, zero chicken'. Below a beaming Colonel Sanders were the words 'Vegan done good', along with 'From the home of the chicken comes a burger that isn't'. Never mind the Eagle, it seemed clear that the *vegan* had landed.

Food culture and the place of meat within it is changing, just as the dominance of dairy is being challenged by plant-based competitors using oats, almonds, soya or coconut. Whatever dairy can do, plant-based milks can do too, and they seem to be elbowing their way to ever more space on the supermarket shelves.

Not so long ago, any mention of veganism would have prompted quizzical looks and confusion. Thirty years ago, some in the media saw the Vegan Society as the 'flat Earthers of food', but things are changing fast – and just in the nick of time.

Time's running out

There is a growing awareness of the environmental impact of meat and dairy. More people realise that factory-farmed meat and dairy are responsible for devouring nearly half the world's grain harvest and most of its soya – enough to feed more than half of all humanity on the planet. We are being mightily short-changed, and the cost doesn't end there. Factory farming causes deforestation, climate change, pollution and destruction of wildlife, not to mention animal cruelty. Whichever way you look at it, there are plenty of reasons to consider eating less meat and dairy.

Like the goats taking over the town of Llandudno, veganism had burst onto the high street. But while the goats would likely soon be off the streets and back on the mountain top, veganism looks set to stay. Together with ending factory farming, eating plant-based meals is one of the biggest things we can do to ensure that we leave our planet in a decent state for our children. Our food habits could have a big bearing on who gets to inherit the Earth.

PLANT PROTEIN REDEFINED

Downtown Decatur, Georgia, and the late-summer heat was quick-drying puddles on streets lashed by the tail of Hurricane Harvey. White marquees had popped up all around. Behind roadblocks, police cars had blue lights flicking like in the movies. There was a real sense of drama yet no cause for alarm. This city had been taken over by America's biggest independent literary festival and I was a guest author. However, my excitement was less about how the audience might react to my book, and more about what I was about to eat. I was going to get my first taste of a new generation of plant-based burger that was said to taste just like meat.

Leah Garcés, my good friend and colleague in Decatur, and her husband Ben were busy preparing Beyond Meat burgers on the barbecue. They were having a party. Adults were popping cider amidst the frenetic play of children.

The centrepiece of the occasion came vacuum-packed from Whole Foods Market. I'd heard so much about these burger patties; entirely plant-based with no soya, gluten or GMOs. Made from pea protein, canola, coconut oil and yeast extract, uncooked, they were salmon-pink and looked distinctly like meat.

Out in their leafy suburban garden, Ben was busy griddling the burgers. He pressed them with a spatula; they dripped and sizzled, flames flickering furiously. They bore no resemblance to old-school veggie burgers, so often looking apologetic, speckled with tiny chunks of vegetable and standing out at a barbecued meat feast like a sore thumb. No, these were imposing, glistening, now striped with charred barbecue gullies. Smelling every bit as meaty as they looked.

So, what did I make of my first of these new-generation plant-based burgers? Full-flavoured, juicy, rich without being overpowering, and with no ambiguity about the taste: big and meaty. However, the 'gristly' bits, designed to give meat-eaters a reassuring texture, took a bit of getting used to.

But did I finish it? I had three.

Joanna Lumley bites

Just weeks later, I hosted London's first ever public tasting of a Beyond Meat burger, with Compassion in World Farming's patron Joanna Lumley taking the first bite. It was part of a celebration of new food technologies at our Extinction conference, where experts from thirty countries gathered to discuss factory farming and runaway meat production that was driving wildlife declines.

Among the leaders of the new plant-based food revolution at the event was Seth Goldman, the executive chair of Beyond Meat. I asked him how rapidly his plant-based burgers were developing. 'The burger just gets better and better', he said, telling me that fifty of his company's three hundred staff are scientists who constantly work to improve the product's taste and texture. The goal is for it to be indistinguishable from meat and without the negative impact.

Traditionally, most veggie burgers have been little more than vegetables mashed together. Beyond Meat started by breaking a hamburger down into its constituent parts and working out how to replicate them. But its secret goes beyond simple ingredients. 'What's really the magic of meat is the way the fats and the proteins are stitched together, and we can use heat and water and pressure and cooling to bring them together in such a way that they replicate the sensory experience of meat', Goldman said.

The other big difference is not so much the product itself, but where it is appears in shops: Goldman's company was insistent that its burgers be located not in the vegetarian food section but in with the meat. 'We want to reach the much larger audience', he said, 'the 95 per cent of people who don't self-identify as vegan and vegetarian. And we know that these people don't buy their protein in the freezer section, so we've got to bring our product to the meat counter.'

He sees new-generation plant-based burgers as the 'seamless replacement' for animal-based protein. 'With no cholesterol and half the saturated fat of a burger, there is a great health impact … On top of that, there's a transformational environmental impact. Our product can use so much less land and so much less water than it takes to feed a cow', he told me.

Goldman's company is just one of those helping to transform how the world sees protein. We are moving from a situation where veggie burgers tasted like cardboard, or worse, to one where they give meat a serious run for its money. In a world with more people and limited planetary resources, one of greatest challenges we face will be the need for more efficient proteins. And replacing the impact of factory-farmed meat with something that has comparable taste could be the key ingredient.

LOSING THE TASTE FOR MEAT?

'The World Is Finally Losing Its Taste for Meat,' claimed the Bloomberg headline.[24] It was the latest in a long line of articles suggesting a dramatic trend in people cutting down on meat. Gallup found 'Nearly One in Four in US Have Cut Back on Eating Meat',[25] while the *Guardian* claimed that a 'third of Britons have stopped or reduced eating meat'.[26] New words entered the public lexicon, with terms such as 'flexitarian' and 'reducetarian', denoting the degree to which people had cut back. The *New York Times* published the paradoxical-sounding 'Meat-Lover's Guide to Eating Less Meat'.[27] Whatever people thought about it, it seemed clear that food culture was shifting to embrace the idea that meatless meals could be cool.

Something in the collective consciousness had swung towards a recognition that eating too much meat was bad, for reasons that included considerations of climate, health and animal welfare. Celebrities such as Arnold Schwarzenegger, Natalie Portman and Al Gore encouraged people to eat less meat to save the planet.[28]

When asked how young men could achieve a body like the cyborg assassin in *The Terminator* without eating steak, Schwarzenegger said that many successful bodybuilders avoided meat. 'You can get your protein many different ways.'[29]

Eating less meat or giving it up altogether had become part of mainstream conversation. Where they had previously been seen as the domain of cranks, faddists and extremists, 'plant-forward' diets had shaken off their taboo, and food companies and restaurants were keen to cash in. Products suddenly became adorned with declarations of being 'vegan-friendly'. Menus started to feature vegan sections and meat-free choices, brandished as badges of honour.

Even the word vegan was embraced by society as a recognised 'food tribe', a term to denote like-minded foodies with distinct preferences. The consumer marketing agency SIVO Insights placed veganism alongside other tribes such as gluten-free, sugar-free and paleo, and an article on its website revealed the way many food companies looked at meat-free trends: 'As marketers, it's our responsibility to be educated on consumers' unique food preferences so that we are able to stay ahead of consumer demand and then anticipate and meet their needs.'[30]

For many companies, meat-free diners were seen as an audience to be either catered for or lost to competitors. Meat was seriously on the wane, or at least that's how it looked, prompting the *Financial Times* to ask, 'Have we reached "peak meat"?'[31] In terms of the expansion of plant-based eating, the free market seemed to be showing it could tackle the environmental threats posed by eating too much meat. Consumer choice had surely won the day. These perceptions were further reinforced by a UN FAO report in 2020, which forecast a decline in meat production for the second year running. For the first time in decades, global meat production was going backwards, albeit by just 1 or 2 per cent.[32]

While some commentators saw this as evidence of the end of humanity's 10,000-year love affair with farmed meat, the reality turned out to be quite different: the same report made it clear that much of the contraction in meat production was down to a sharp fall in pig numbers due to outbreaks of disease in Asia, as well as Covid-related disruption of the meat market.[33] In China, African swine fever was estimated to have infected between a third and a half of the country's pigs, with millions culled.[34] So the Bloomberg headline hadn't yet been matched by statistical evidence or credible forecasts – at least, not unless one conflated the word 'meat' with 'beef'.

MEAT OR BEEF?

At the time of writing, the data points to people being more open than ever to trying plant-based meat alternatives, while also eating less beef but more chicken and fish. Sales of plant-based alternatives have grown quickly, but signs of a decline in overall meat consumption remain elusive. In March 2020, sales of fresh meat alternatives in the US were up by between 300 and 400 per cent compared to 2019.[35] However, their market share continued to be dwarfed by meat. Plant-based alternatives were worth $12 billion in 2019 and were forecast to reach $28 billion in 2025,[36] but the global meat market remained nearly eighty times bigger, at $946 billion.[37]

What the data showed was that beef had taken a hit as a result of health and environmental messages about too much red meat causing cancer and being bad for the climate. Consumption of beef declined globally by less than 1 per cent,[38] with the void being filled by chicken and fish. Between 2014 and 2018, global poultry meat consumption increased by 4 per cent and fish consumption rose by 2 per cent.[39] A survey by Public Health England, looking at the period 2008 to 2017, found a downward trend in people eating red meat (beef, lamb, pork, sausages and burgers) by about 1 per cent per year. Chicken was increasing, particularly among teenagers and those over sixty-five. Teenagers were also eating more fish. However, the overall picture for meat consumption was one of 'little change'.[40]

Beef had become vilified, and few people were drawing a distinction between meat from planet-destroying feedlots and the more sustainable pasture-fed variety. Thousands of cattle standing together in a single feedlot were viewed in the same way as a few cows roaming freely on grass; as a result, the benefits of keeping cattle in small numbers as part of regenerative agriculture were getting lost.

Even fewer people noticed that replacing beef with chicken and fish means more animals are factory-farmed to produce the same amount of meat. Cows, being so much bigger than chickens, produce much more meat. Consequently, switching from beef to chicken multiplies the number of animals involved in the system by a factor of 160,[41] and the multiplication factor for fish can be even worse. So one cow on grass may be replaced by 160 chickens, most likely reared on factory farms. The majority of the world's chickens are farmed intensively in crowded conditions, pushed to grow super-fast and killed at six weeks old. Half of all fish produced are now farmed, many intensively. In the cruelty stakes, a straight switch from beef to chicken or fish makes matters worse rather than better.

While data shows diets switching from beef to chicken, it also shows that overall consumption of animal products, be they meat, dairy or eggs, is increasing. Between 2014 and 2018, UN global food supply figures showed overall consumption of meat (including fish) per person rose by 4 per cent, while the EU and the US registered upticks of 2 and 6 per cent respectively.[42] In the US, despite the profusion of plant-based alternatives, people were eating more meat than ever.[43] There are notable exceptions; in Sweden, for example, there was a 4.5 per cent reduction in meat consumption in 2020 compared to the previous year.[44]

Yet despite warnings that we should rein in our consumption of meat, 'peak meat' for much of the world could still be ahead of us. Animal-based meat production will only decrease if levels of consumption per person decrease and human population size remains the same: neither looks likely in the short term.

Peak meat

According to studies, global meat consumption is predicted to grow over the coming decade.[45] The Organisation for Economic

Co-operation and Development forecasts a 1 per cent increase in meat consumption per person per year globally to 2030.[46] Although this is half the growth rate of the previous decade, it nevertheless means that the average global citizen is set to eat more meat than ever. A study in the *Lancet* was even less optimistic, predicting that average meat consumption per person globally will increase by more than a quarter by 2050, with already big meat-eating regions like the US and EU continuing to see average consumption increase by about a further tenth. If we take population growth into account, our global appetite for meat could increase by 78 per cent by the middle of the century.[47]

The Earth's population is set to increase by 2 billion people in the coming decades, and rising animal production correlates closely with population growth. For every billion people, 10 billion land animals are farmed every year, a ratio that looks set to increase as diets shift towards chicken. Global meat consumption, of course, is far from uniform, with countries like the US, Australia and the EU eating much more than the world average. Yet as the world's population increases, so does the global demand for meat.

At the heart of growing unease with spiralling meat consumption lies the fact that animal products are responsible for the majority of agriculture's greenhouse gas emissions.[48] If the world carries on eating meat as it is, our food could blow our chances of keeping global warming in check.[49] Simply shifting from beef to chicken won't help much, as chickens eat vast quantities of cereals and soya, thus contributing to deforestation, carbon-emitting soil degradation and nitrous oxide emissions, the most aggressive greenhouse gas, from fertiliser use.

As governments start to take the climate emergency seriously, the farming of so many animals presents a major obstacle to our attempts at curbing global warming. There are technical things that farms can do to reduce emissions from animal production,[50] but this would only make minor inroads into a much bigger problem.[51] For the kind of results needed to combat climate change, the real opportunity lies in tackling meat consumption.

Cut meat by half

A 2018 report by the Rise Foundation found that the EU needs to cut its livestock-related emissions by three-quarters by 2050 to meet climate change targets. Achieving that will require drastic changes in the way we view meat in food culture – looking at it less as an everyday staple and more as a treat. As it is, the EU predicts that meat consumption over the coming decade will decline by just 2 per cent,[52] nowhere near enough to hit climate targets.

While consensus builds that a shift away from our reliance on meat is urgently needed, the reality is very different. Turning plant-based eating from food trend to planetary lifesaver will take more than fine words and optimistic statements. On current data, simply leaving it down to the marketplace is a bad idea.

Policy action

If we are to reach 'peak meat' and see genuine reductions in meat consumption, we need concerted action from policymakers in governments, the UN and leading corporations. The influential think tank Chatham House called on European governments to 'take stock' and include meat consumption and promotion of alternatives among their 'policy priorities'. It concluded, 'In order to meet its climate change commitments, the EU will need to change European eating patterns, including a reduction in meat consumption.'[53]

In other words, if we are to avoid environmental disaster, policymakers must grab the plant-based bull by the horns and reduce meat consumption fast. But doing that will require governments to send clear policy signals to citizens; as Chatham House puts it, 'to prompt shifts in consumer food choices, away from the most resource-intensive meat products and towards more sustainable alternatives'.[54] They will need to clear a path for favourable regulatory, labelling and marketing rules, allowing meat alternatives to be desirable, affordable and accessible. It will take public money to move away from supporting intensive meat production and towards alternative sources of protein. And above all, it will need governments to resist lobbying pressures from the meat and agribusiness industries and to stand up for the best interests of their citizens.

However governments manage to shift diets in the future, trading beef for chicken isn't the same as cutting back on meat. But in the minds of some journalists, environmentalists and consumers, there is a danger the two have become confused. The cruelty and environmental credentials of intensively farmed chicken is every bit as shocking as feedlot beef.

Meat alternatives, whether they are plant-based or grown from stem cells, offer the prospect of sating our hunger for meat, without the climate-busting consequences. Eating less but better meat has become a common mantra, even if it hasn't yet registered in statistics. Being careful to eat only animal products that are regenerative, pasture-fed, free range or organic is crucial. By avoiding factory-farmed meat, we can make sure reductions are gleaned from the intensive sector.

Thus, it seems that we can find a workable way to reduce our meat consumption by half. Making a distinction between factory-farmed and regenerative meat will be key to ensuring that reductions add up to something more sustainable. And as we retune our diets to eat more plants and less meat, the bottom line is that 'meat' doesn't equal 'beef'; confusing the two could well turn out to be the cruellest miscalculation ever made.

MEAT WITHOUT ANIMALS: CULTURED MEAT

Streaking across the night sky, the International Space Station was undertaking an extraordinary mission: to produce the first beef in space. Nearly 250 miles above the Earth, Oleg Skripochka was pictured wearing a white T-shirt and clutching an orange box. He looked remarkably relaxed for someone in the process of doing something so groundbreaking. A cow was conspicuously absent; instead, cells from a cow had been blasted into orbit aboard the Soyuz MS-15 space rocket. Skripochka and his cosmonaut colleagues mixed them with a nutrient soup and added the mixture into a 3D bioprinter. They were about to 'print' the first 'cultivated beef steak' ever to be produced in space.[55]

It was 26 September 2019, and the spacecraft was hurtling at 15,000 miles an hour,[56] while growing a tiny piece of cultured steak that proved meat no longer needed to come from an animal. Although the breakthrough came in space, its implications were much greater for life on Earth. The technology behind that first space-steak promises to reshape food on the planet we call home.

'We are proving that cultivated meat can be produced anytime, anywhere, in any condition,' said Didier Toubia, co-founder and chief executive of Aleph Farms, the Tel Aviv-based company behind the breakthrough. Referring to the fact that producing meat from animals is heavy on resources, Toubia said, 'In space, we don't have 10,000 or 15,000 litres of water available to produce one kilogram of beef … This joint experiment marks a significant first step towards achieving our vision to ensure food security for generations to come, while preserving our natural resources.'[57]

Aleph Farms' effort to produce 'slaughter-free meat in space' came at around the same time as urgent warnings that the world will face catastrophic climate change if our taste for meat isn't curbed. Earlier that year, a report adopted by the Intergovernmental Panel on Climate Change pointed to high-meat diets from intensively farmed livestock as undermining efforts to keep the world at a liveable temperature and cited cultured meat among the answers.[58] A longstanding study had shown cultured meat production to be 80 to 95 per cent lower in greenhouse gas emissions and 98 per cent lower in land use than conventionally produced meat products.[59] This was reinforced by a 2021 lifecycle analysis study which found cultured meat produced using renewable energy to have up to 92 per cent lower climate impact and needing as much as 95 per cent less land than meat from farmed animals.[60]

A spokesperson for Aleph Farms reinforced the message that the experiment to grow cultured meat in space was about saving the Earth: 'Our planet is on fire and we have no other one today. Our primary goal is to make sure it remains the same blue planet we know also with our next generations.'[61]

Although the first in space, Aleph Farms was always unlikely to be the first company to market cultured meat, due to the work needed to properly replicate the structure of a steak. That crown would more likely be claimed by a 'first-wave' company geared towards producing cultured meat with less structure – like mince, for burgers, sausages or nuggets.

Cultured meat was very much in its infancy in 2020, but still, big companies like Tyson could see its potential and were honing-in on developers of 'first-wave' products. KFC, one of the world's biggest fast-food chains, partnered with the Russian company behind the 3D bioprinter used in the space-steak mission, with the aim of creating the world's first cell-based chicken nuggets.[62] The company said in a statement that what it called 'crafted meat products' were part of KFC's 'restaurant of the future concept'.[63]

While Aleph Farms' steak was seen as the second wave after burgers and nuggets, companies were already wondering whether replacements for fish and crustacea would be part of a third wave to reach the market. Among the leaders were San Francisco-based Finless Foods, who produced real fish from stem cells. To do this, they multiplied cells in a fish-derived serum, before structuring them into fillets and steaks. While the finished product wasn't yet animal-free, it was claimed to be almost entirely cruelty-free, avoiding contaminants, overfishing and ocean pollution. With bluefin tuna first on their product list, the team planned to expand into cell-culture versions of other high-value fish species.[64]

Perhaps the fourth wave is being defined by Sandhya Sriram, a stem cell scientist who sees the ability to grow cultured meat in our homes as the next frontier. She envisions people having a small bioreactor – something a bit like a pressure cooker – that can maintain the right temperature and conditions for cells to grow into meat. Sriram is co-founder and chief executive officer of Shiok Meats, the Singapore-based company growing cell-based shrimp in a laboratory. In an interview with Channel News Asia, she said, 'It's much like making beer or wine at home, or even baking a piece of bread.' She estimated that the technology could be widely used in as little as ten years.[65]

How it works

Unlike plant-based 'meats' headlined by the Impossible Burger, Beyond Meat or Quorn that draw their protein from sources such as soya and peas, cultured meat at the cellular level is undisputedly meat. It is grown in a culture using the same kind of cells that would otherwise make up an animal. While the scientists, investors and regulators involved in the technology have yet to alight on a unified term, 'cultured meat' remains an accurate description. The cells used to start the process are harvested from a living animal in a harmless biopsy. These stem cells from the fat or muscle of an animal are placed in a culture medium – a nutrient-rich soup – that allows them to grow in a bioreactor similar to those used for fermenting beer and yogurt. No GMO is required – the cells do just what comes naturally: multiply. And to staggering effect: a single sample from a cow can produce 80,000 quarter pounders.[66]

Millionaire's hamburger

Cultured meat had its first public taste test in 2013. It was lunchtime in London and two hundred journalists and academics packed into an auditorium to watch nervous panellists eat a burger that was completely unique. At the centre of the crowd was Professor Mark Post from the University of Maastricht, his hands poised to reveal a culinary creation that had cost $280,000 to produce. The event was more reminiscent of a TV food show than a scientific announcement. The audience watched with bated breath as the presenter Nina Hossain asked the professor to lift the lid on his creation. Without a modicum of ceremony, Post removed the silver platter to reveal a glass dish of raw, pink minced beef. To be honest, it looked rather ordinary, which was the point really. It was then cooked for the waiting food critics, who were asked whether this stem cell meat tasted like a regular burger. 'I was expecting the texture to be more soft,' said Hanni Rützler of the Future Food Studio, who researches food trends and was the first to taste the cultured burger. She looked mildly sceptical. 'There is quite some intense taste; it's close to meat, but it's not that juicy. But the consistency is perfect,' she added.[67]

Mark Post's original burger, funded by Sergey Brin, the co-founder of Google,[68] was way beyond the budget of the average consumer – a pound of cultured meat was said to have cost more than a million dollars. However, by 2020, the cost had dropped to about $50 per pound,[69] still too expensive for the mass market but heading in the right direction. By then, Post had become known as the 'founding father' of cultured meat,[70] while his company, Mosa Meat, had been joined by a plethora of start-ups, all vying to be first to market. Some thirty-five companies were listed in 2020 as producing cultured meat – ten were in the States, five in Israel, and the rest were in Argentina, Australia, Belgium, Canada, France, Hong Kong, India, Japan, the Netherlands, Russia, Singapore, Spain, Turkey and the UK.[71] A similar number were working on component parts of the process, focusing on better growth mediums and the other 'picks and shovels' needed to make this new industry fly.[72] Not surprisingly, as the twenty-first century rumbled into its third decade, still more people had ventured into outer space than had tasted cultured meat.

Barriers

As with any new technology, getting the price down and scaling up are two of the biggest hurdles. There is also the regulatory vacuum; current legislation never foresaw a world where meat would come from anything other than an animal. Major investors might have been put off by the thought that their carefully crafted invention might turn out to be illegal. As it is, the conventional meat lobby is marshalling against the new competitor. In the US, state legislation in Missouri forbids cultured meat from being labelled 'meat', and other states have been inspired to follow suit.[73]

Chase Purdy, the journalist and author of *Billion Dollar Burger*, believes that the biggest barrier to cultured meat entering the marketplace is not the technology itself, but regulation. 'The technology is ready. The science is there, the science has been there for a while. It's really all about governments around the world figuring out how to regulate these products.'[74]

On the ethical side, the use of serum drawn from the blood of a cow's foetus, used in the cell-growing medium, has long been a stumbling block for what some in the industry call 'slaughter-free meat'. Foetal bovine serum is also expensive, which presents a cost barrier. Scaling up requires something cheaper. Researchers have developed media from a variety of other sources, including plants and microorganisms.[75] However, it is hard to determine their success because companies keep a tight lid on trade secrets.

Market entry

Cultured meat could be one of the food industry's biggest disrupters. Chase Purdy thinks it won't be long before meat from a bioreactor is sitting next to meat from a cow. 'If cell-cultured meat can be about the same price as the meat we buy today, I think that if it looks the same, if it tastes the same, it will be just as common to see it in a grocery store as conventional meat.'[76]

When that day will come, however, has long been a subject of conjecture. According to a 2019 article by *TechRound*, companies were claiming that they would be market-ready between 2021 and 2023.[77] However, New Harvest, an institute funding research in the sector, is more circumspect. 'We do not have any formal predictions or promises regarding when products may become available commercially,' it said on its website.[78] Citing scientific unknowns and regulatory hurdles that make it 'very difficult' to commit to time frames, Mark Post's Mosa Meat would only go so far as to predict market entry 'in the next few years'. Either way, the company sees initial market entry being small-scale, with wide availability in supermarkets being 'several years beyond that'.[79] A survey of cultured meat companies carried out by the *Daily Telegraph* in 2020 found that most thought market entry was between five and ten years away.[80]

Consumer acceptance

Beyond price, scale and regulation, the final hurdle for cultured meat will be consumer acceptance. Research by the strategic marketing firm Charleston Orwig suggests that more than half of respondents in the US would be open to trying cultured meat, even if was labelled with the negative term 'lab-grown meat'. The study found an emerging awareness, especially among the young, that new technologies will inevitably become part of the food system.[81]

A big part of the success or failure of cultured meat will depend on how it is marketed. The marketing of conventional meat relies on people being comfortable with the image of where their meat comes from: an image of happy cows grazing on grass at Old Macdonald's farm, while a few chickens peck around in the field. The reality is very different, with most animals living short, miserable lives on factory farms before being dragged off to be slaughtered in horrendous conditions that resemble hell on Earth.

Factory-farmed meat, which accounts for the overwhelming majority, relies for its sales on being cheap and anonymous. It tends to be sold with little labelling beyond the word 'fresh'. Given a straight choice and honest labelling, how many people would really choose 'factory-farmed meat' over 'cultured meat'?

Whether people will accept cultured meat comes down to how it is labelled and their perception of 'natural'. Today, very little of our society – cars, electricity, computers, cities, factory farms – is natural. As for whether cultured meat itself should be considered natural, Chase Purdy raises an interesting point when he asks, 'What is cell-cultured meat other than man's attempt to recreate something that nature has already given us?'[82]

The real choice facing humanity then is whether we restore the previously held contract with the land, with the soil, and produce food in harmony with nature. This would involve dispensing with much of the meat currently on supermarket shelves, not to mention reducing the amount of animal-sourced foods we eat. It would require us to dispense with industrial agriculture and restore farmed animals to the land as free-roaming creatures alongside resurgent wildlife. Or do we let nature survive by taking food production out of the countryside altogether, taking it into cities where most consumers in the world now live? This latter scenario is the prospect offered by cultured meat.

As with so many problems in life, the answer perhaps lies in a blend of solutions: creating mass-market meat using bioreactors in much the same way as craft beer, while restoring the countryside to a thriving landscape through regenerative farming geared around feeding people. After all, this would transform the food system into one focused on growing food for people rather than *feed* for factory-farmed animals. It would also boost soil fertility, something needed for a plant-based future as well as an omnivorous one. With pressure for change building in the face of the climate crisis and the collapse of nature, impetus is growing for a major rethink of how our relationship with meat works. While drastically reducing our consumption is increasingly seen as part of the answer, our appetite for meat continues to rise. Cultured meat could well offer a lifeline and buy precious time in the race to save the planet.

That blend of approaches – cultured meat sitting beside regenerative farming – was a theme echoed by the man behind the first cultured beef steak in space. Before he launched those stem cells inside a rocket ship, Didier Toubia from Aleph Farms said his company was not aiming to replace traditionally raised, grass-fed cattle. 'We are not against traditional agriculture. The main issue today is with intensive, factory farming facilities, which are very inefficient and very polluting and have lost the relationship to the animal.'[83] When I think back to that space mission, I am struck by how its creators saw their new invention sitting beside the best of what already exists, preserving life on Earth for generations to come. And this seems to me to be the cultured thing to do.

FROM FIELD TO FERMENTER

When it comes to the future of food, a great deal of media attention has focused on cultured meat, perhaps because, being actual meat, the product seems relatable to the mind's eye; however, there is a more immediate transformation of the food system approaching: precision fermentation.

Precision fermentation is based on the same symbiotic relationship formed over millions of years between the cow and the microbes in her gut, only without the cow. It's based on the idea that microbes can be trained to produce specific building blocks of food without any need for an animal.

At its simplest, food is made up of packages of nutrients, whether they are proteins, fats, carbohydrates, vitamins or minerals. Precision fermentation allows for those constituent parts to be built to precise specifications. There's talk of 'food as software', a technique where food engineers can use 'molecular cookbooks' to design food products in much the same way as software developers design apps. And compared to the industrial production of food from cattle, the efficiency savings of precision fermentation could be huge: it is reported to be a hundred times more land-efficient, up to twenty-five times more efficient at converting feed into food and ten times more water-efficient.[84]

The key to the next generation of food production is decoupling microbes from farmed animals and going straight to bacteria and fungi, where the real work gets done. Cows were domesticated from wild aurochs between 8,000 and 10,000 years ago for their meat, milk, hides and their ability to pull a plough. The powerhouse of the cow is her four-chambered stomach, which acts as a fermentation vat.[85] Trillions of microbes turn tough plant fibres into digestible nutrients that fuel the animal through her life, at the end of which she is slaughtered and turned into cuts of protein.

Nature intended for those microbes to help cows and other ruminant animals to feed on grass, herbs and leaves. The largest of their four stomachs, the rumen, acts as a holding tank from which they can regurgitate and rechew food – 'chewing the cud' – meaning they don't have to chew when they are in the vulnerable head-down position while grazing. Industrialisation has led to many ruminants being taken out of fields and fed grain instead of grass, but as we have already seen, the transaction is pitifully inefficient. Cattle for beef waste up to 96 per cent of the food value of grain in converting it to meat – those microbes work hard for little return.

Cutting-edge food producers are beginning to sidestep the cow, taking the microbes out and setting them to work directly – and far more efficiently – in modern fermentation tanks. It's what some people are calling the 'second domestication' of plants and animals: producing food by domesticating microbes directly rather than bothering with the animals that traditionally harboured them.

It's an age-old process (think bread and wine) with a modern twist: cylindrical stainless-steel vessels, where ingredients and microbes coexist in a controlled environment. The technology builds on the work of nineteenth-century scientists like Louis Pasteur, who by peering through their microscopes, learned to control and manipulate microorganisms.

Precision fermentation offers tremendous scope to transform the world of protein, with new types and availability that were hitherto unimaginable. It builds on the fermentation process used for millennia, where microbial cultures have been used to preserve food, create alcoholic drinks and produce foods ranging from yogurt to kimchi and tempeh. More recently, the technique has been developed into 'biomass' fermentation, where fast-growing microorganisms such as algae and fungi are used to produce protein in vast quantities, the organisms themselves becoming the food or the main ingredient. Many microorganisms are naturally high in protein and replicate extraordinarily fast. Unlike chickens, pigs and cattle, which can take weeks, months or years to mature, microorganisms can double their numbers in hours or minutes, meaning that great quantities of food can be produced in relatively short periods of time. The pioneering UK food brand Quorn has been leading the way for decades, creating protein-rich plant foods by fermenting cereals with a natural fungus found in the soil.[86] In America, Meati Foods uses biomass fermentation for its 'mycelium', a microscopic thread-like structure that is used in the company's plant-based steaks.[87]

Precision fermentation takes things up a few notches, using specific strains of microbes, often altered or 'programmed', to produce precise proteins or other ingredients. Compared to the few dozen animal species and several hundred plant species that humans routinely eat, microbes offer endless opportunity for food applications. As many as one trillion species of microorganisms are thought to exist on Earth.[88] Instead of rearing an animal to get protein from the meat, microorganisms can produce individual nutrients directly. Furthermore, food can be made of specific nutrients and to exact specifications, avoiding the painful business of rearing and killing animals to access them. And in shifting food production to the molecular level, the number of potential nutrients expands enormously; no longer constrained by what is available from the plant and animal kingdom, food moves from being a process of extraction to one of creation.[89]

If all of this seems far out, think again; products that are either scarce in nature or uneconomic to farm, like natural vanilla, orange flavouring, sweeteners and vitamins, are already produced directly from microbes.[90]

Recent years have seen a veritable explosion of companies producing alternative proteins from fermentation. While cultured meat has been getting all the press, in 2019 fermentation received three and half times as much investment. Unlike cultured meat, which has been delayed reaching the market by regulatory red tape, fermentation is already a commercial reality. The basic technology is long established but the advent of 'precision biology' combined with fermentation has opened the door to a world of possibilities for the future of protein.[91]

The approach, which often uses genetic modification to manufacture specific animal and plant proteins, fats and other molecules, won't be to everyone's taste. But when it comes to trade-offs, it is surely better to use genetically modified microbes in fermentation than to grow GM corn or soya on vast acreages of cropland to feed factory-farmed animals kept in abject misery. Like it or not, genetic modification is a major element of industrial livestock production. Precision fermentation could be part of a protein revolution that leads to a future without factory farming. One way or another, a major shake-up of the food industry is on the cards.

To some degree, fermentation has already caused disruption. Fifty years ago, rennet, an essential ingredient for cheesemaking, came from the stomachs of veal calves. That was until a campaign by the animal welfare movement about the plight of calves kept in darkened crates or killed at birth made veal a dirty word. Calf rennet became expensive and when the industry looked for alternatives, precision fermentation came to the rescue. Pure chymosin, the active ingredient in rennet, could be produced cheaply and efficiently through fermentation. Today, fermented chymosin is used to produce more than 90 per cent of cheese in the US.[92]

As yet, the alternative protein industry has barely scratched the surface of what is possible and stands to revolutionise the creation of protein. As the cost of fermentation plummets and our ability to programme microbes to set specifications increases, the stage is set for these new techniques to supplant industrial animal agriculture, and according to scientists, the possibilities are endless. The same process could also be used to replicate proteins from extinct plants or animals. Leather or meat could be developed from mammoths or whales, and all without harming animals, farmed or wild. Woolly mammoth burger, anyone?

BIOMASS FERMENTATION

In a setting said to have inspired Aldous Huxley's *Brave New World*, two huge stainless-steel cylinders rose up out of a warehouse, with a mass of pipes behind them. Rows of people peered at computer screens as they oversaw the fermentation of wheat into protein, all thanks to fungi.

I had come to Billingham in the north-east of England to find out about a protein source that was giving meat a run for its money. What I was witnessing was the production of a food from fermentation that is already firmly established on supermarket shelves in Britain and around the world. A tiny member of the fungi family called mycoprotein grows explosively in each fermentation chamber, producing enough for nearly 100,000 burgers a day.

My visit came in the wake of a report by researchers at Murdoch University in Australia, which suggested that artificial meat could turn conventional meat into a premium luxury as the world's population grows and livestock production fails to keep pace with demand. The study suggested that meat producers will need to find solutions to animal welfare, health and sustainability issues 'in the face of competition from emerging non-traditional meat and protein products'.[93]

The food scientist Tim Finnigan worked for Quorn Foods for more than twenty-five years. He told me how the product was aimed at the 'flexitarian' market – those people who want to eat meat but are cutting down in favour of plant-based alternatives. Often, the motivation is health or the environment. 'And that's where Quorn is really helpful in transitioning, because it's familiar,' Finnigan said. 'It's not asking you to do anything weird or different, you can still have chilli, spaghetti bolognese, all your usual foods – and just as good in the majority of cases, I would argue – but without having to use meat.'

The discovery of mycoprotein in the 1960s was the culmination of a worldwide quest by Lord Rank of the bread manufacturer Rank, Hovis and McDougall, aimed at finding a new protein source. It is now the mainstay of a £230 million food business called Quorn.[94] 'What they were looking for was a microorganism, one that would convert the then plentiful carbohydrate into the less plentiful protein,' Finnigan explained. Three thousand soil samples later, it was found in a compost heap in Marlow, Buckinghamshire.

So what are mycoprotein's credentials? Well, it's better at converting grain into edible protein than farm animals, with a smaller impact on scarce resources. It takes between twelve and twenty-four kilograms of cereals to produce the average kilogram of beef, and even chicken takes two to four kilograms. Quorn is different. According to the company's information, a kilogram of mycoprotein requires just two kilograms of wheat, converting the grain's carbohydrate into protein without the need for an animal. What's more, a greater amount of protein is available at the end. 'You don't have to use land to grow protein to feed to another animal that then reduces the yield of protein in the final product', Finnigan said. 'You actually get more protein than you started with, which is the opposite of the livestock story.'

Quorn's greenhouse gas emissions have been found by the Carbon Trust to be thirteen times lower than beef and up to 4 times lower than chicken.[95] The product was given a boost by the European horsemeat scandal in 2013, and the company's orange logo can now be found on supermarket shelves throughout Britain, Belgium, Scandinavia, Australia and elsewhere. After all, if there's no meat in the product, there's no chance of finding horsemeat in it.

So is Quorn one of the 'emerging' alternatives that are threatening to put meat in its place? If the shelves of my local supermarkets are anything to go by, it seems to be one of the current front-runners. Even our tiny village shop has Quorn sausages in the chiller cabinet. By virtue of its fermenting process, mycoprotein makes possible food with a low-environmental impact that might even be manufactured in cities. But how do people feel about it? Some say a local taxi driver is a good gauge of popular opinion, so on the way to the company's headquarters, dubbed the 'Quorn Exchange', I asked the driver what he thought of food made out of mycoprotein. 'It's a good idea', he said. 'It can feed the world, can't it?'

CELL CULTURE AND THE CONSCIOUS CARNIVORE

Never had I felt so excited about a dinner date, nor more looked forward to eating chicken. I felt giddy as my head raced to think how I would fly out to the States without breaking Covid-19 restrictions. I had just been invited by an impassioned entrepreneur to join him in California for the first public tasting of his company's cultured chicken meat. It felt like an invitation to taste the future.

'It would be beautiful if you could make it', said Josh Tetrick, co-founder and CEO of the food start-up Eat Just, Inc., based in California. When I spoke to him over Zoom, he was in the mountains of Montana. Looking relaxed and unshaven in a blue T-shirt and bright red baseball cap, he told me all about cultured meat. I was particularly interested in how it related to plant-based alternatives, fermentation and regenerative farming. A law school graduate, forty-year-old Tetrick once worked for the Liberian government supporting reform of the country's investment laws. He also spent several years working with non-profits and the UN in sub-Saharan Africa.[96] His work included teaching street kids and encouraging child prostitutes back into schools. It was during his time in Africa that he started to worry about the global food system. He saw children who were suffering from a deficiency in micronutrients. Shocked by this hidden hunger, he resolved to do something about it.

What became obvious during our conversation was that he was following his calling to change the world. To him, running a successful company was more about making change than making money. And unlike many company bosses, when I asked him about the profusion of other companies developing cultured meat, he told me he welcomed the competition. In fact, he said, he'd be worried if more cultured meat companies weren't emerging. After all, the challenges around factory farming were getting more pressing every day.

'We have an urgency, and every single second we delay is creating more pain and more degradation and taking us further from who we are [as a species] … If you told me right now that some other company will solve this problem in the next handful of years, that more human beings will eat meat that didn't require killing an animal and it had nothing to do with us, sign me up – I'll just go on vacation in the mountains!' he said.

During our conversation, it became obvious that cultured meat was being held back by a regulatory system that didn't know how to deal with cultured meat – when lawmakers had written existing food legislation, no one thought that meat would come from anything other than a slaughtered animal.

Talking in September 2020, Tetrick shared his hope that cultured meat might soon get regulatory approval. Within sixty days of getting the go-ahead, Tetrick reckoned he would be hosting his long-planned celebratory meal. He'd be signalling the sun coming up on a new era for meat with the first public launch of his cultured chicken. He was all geared up. He just needed the law to play catch-up.

Such was his eagerness to get cultured meat in front of the public that Tetrick was no stranger to disappointment. He told me about a moment in 2017 when he flew from San Francisco to Amsterdam to make the first commercial sale of cultured meat. The plan was to do the deal in the European Union ahead of new legislation introduced to restrict the sale of novel foods. After 1 January 2018, new products would have to go through an exhaustive application process to get government approval. If he could sell his product before the new law came in, a case could be made that cultured meat was already in the marketplace and it wouldn't face such stringent restrictions. What could possibly go wrong? Well, Tetrick made it to Amsterdam but was left in the arrivals hall at Schiphol Airport, waiting forlornly beside an empty luggage carousel – an airline error meant that his stem cell meat had never left San Francisco. Many nail-biting hours later, his luggage arrived with the high-tech meat still inside. The sale went ahead, in the nick of time. A restaurant in Zaandam bought just over a pound of cultured meat at a knock-down price. For Tetrick, it was mission accomplished, but the victory was short-lived. The planned first tasting was stifled by regulatory red tape.[97]

Three years later, the regulatory hurdles seemed impenetrable. Once the red tape gets lifted, Tetrick planned to roll out his cultured meat far and wide, with the aim of seeing it 'normalised'. On the journey to normalisation, he saw 'precision' fermentation as an exciting technology that could help bring costs down, making cultured meat affordable and ubiquitous without the need for animal ingredients in the growing medium. That soup commonly contains an active ingredient found in the blood of baby cow foetuses.

'It somewhat defeats the purpose of having meat without killing an animal if the main input needs to be blood from a foetus, right?' Tetrick suggested. Precision fermentation, with its ability to harness the power of microbes, stood out as one of a number of different options that could provide a growing medium without the need for foetal cow blood.

There was also the cost of producing the special soup in the first place. 'We need to figure out a way to make the growing medium to such a low cost so that we can launch it [cultured meat] and be as ubiquitous as Coca-Cola', Tetrick said. Companies were popping up to focus on solving just this problem, and the production of nutrients through precision fermentation could be part of the solution.

What became obvious was the symbiosis developing between cultured meat and other alternative proteins. As the technology developed for each, their overlapping nature seemed likely to make them both cheaper and more viable. Protein of the future will be produced in a range of ways. Some will be traditional – like regenerative farming, with its soil-enhancing, welfare-friendly animal rearing in harmony with nature – while others will be more technologically driven – like the 'modern' foods from cultured meat and precision fermentation.

Precision fermentation could make plant-based products cheaper by producing soya isolate – a key ingredient in some plant-based meat alternatives – without the need to actually grow soya. Through producing cheaper and animal-free ingredients for the nutrient soup used to grow stem cells, it could enable cultured meat to scale up at an affordable price too. When that happens, the inefficiencies of production of today's industrialised agriculture and the realities around cost could mean the number of animals farmed will drop dramatically. The remaining animals could be given decent lives on regenerative farms. So, choosing between plant-based, cultured or fermented protein is an artificial choice; if we are to move on from damaging factory-farmed meat, we'll need all the solutions at our disposal.

The most important thing is that meat from factory-farmed animals should be absent on the menu of the future. There remains a concern that the plant-based meat option on menus will remain just that: an option. For the time being, most restaurants are likely to feel compelled to continue offering 'real' meat. Cultured meat could plug the gap – and being 'real' meat, it could be a game-changer. Restaurants are more likely to drop meat from animal agriculture if they can offer cultured meat. As Tetrick said, 'I look at a world where I don't just want plant-based meat or cultured meat *on* a menu – I want it to be the *only* thing on the menu.'

Regenerative farming, in combination with 'modern' foods like cultured meat and the new generation of fermentation techniques, offers huge scope for turning around the tanker of industrial agriculture. So we shouldn't see it as plant-based versus cultured meat versus fermentation versus regenerative farming – it's all of them together. Just as we wouldn't want the energy sector to become a polarised debate between wind turbines and solar energy, solving the food crisis needs a multiplicity of solutions. If I've learned one lesson from the factory farm era, it's that a 'one-size-fits-all' approach is no way to go. It's a lack of diversity – with intensive farming being seen as 'the' solution – that got us into this mess. And it's a diversity of progressive solutions that will get us out of it.

As for the big breakthrough Tetrick had longed for, it finally came in Singapore in December 2020, where the world's first regulatory approval was granted for the sale of cultured meat. Groups of diners in evening dress and smart-casual attire, including government officials, investors and four environmentally aware children and their teacher, perused the menu based around 'Good Meat cultured chicken'. Slices of breaded chicken pieces were served in bamboo bowls and on granite plates. Three offerings symbolically covered styles from the US, China and Brazil, the three biggest chicken-producing countries of the world. Tables were forested with tall-stemmed glasses. Restaurant walls were alive with moving images of colourful birds' feathers, city lights and shoaling fish. Lots of obligatory iPhone pictures were taken and many smiles were exchanged. There were clean plates; the cultured meat had been roundly polished off.

Then came the moment restaurant-goers normally dread: paying the bill. Only, this time, it would constitute a priceless memento. At 7.23 p.m. Singapore time, the first restaurant bill for cultured meat was settled, with each diner paying $23. History was made.

But after all that, what did it taste like? A twelve-year-old diner gave his opinion: 'This chicken ... it's just chicken, but it's the most amazing thing I've ever seen or ever tasted. It's definitely made me see how small things, like just changing the way we eat, can literally change our entire lives'.[98] His eleven-year-old companion chimed in: 'I'm speechless ... It feels good to have chicken without feeling guilty.'

After decades of getting it wrong, there seems to be a new way bubbling beneath the surface: the birth of new techniques for creating protein, coupled with a rebirth of regenerative farming. So not one single solution, but lots of them – a profusion of possibilities that will transform our attitudes to 'meat'. As we're running out of time to change, solutions that allow people to eat meat without the negative consequences could just save society's bacon.

And, like Tetrick, I sense the need for a big shift if we are to preserve our way of life and our food culture. Which is why I vowed, when Covid restrictions were lifted, to take up his offer to join him for a celebratory meal of cultured meat. Never will chicken have tasted so good.

TOWARDS 2100 – PROTEIN'S NEW DAWN

When history looks back on this period, will it be remembered as an era with the same unfettered imagination shown during the dawn of agriculture? That chapter opened 10,000 years ago on the fertile banks of the Tigris and Euphrates rivers, when nomads became farmers of sheep, goats and later cattle for meat. From that moment, our fate no longer lay in the hunter's luck of the chase but in the health of the soil and our ability to nurture crops and animals on it. And so domestication of plants and animals began, with breeds adapted to our requirements. They became a living larder. A coveted source of protein.

This was Mesopotamia, now Iraq, way back in the Stone Age, where the rich supply of protein from meat and other animal products nourished the very cradle of civilisation. It enabled hunter-gatherers to settle in communities, creating craftsmen and priests, soldiers and scribes. This led to the invention of the wheel, the chariot and the ox-drawn plough.[99] It spawned the dawn of invention and cast in stone our deep-rooted association of protein coming from animals.

In this way, the business of farming animals for meat emerged into a big wide world of few people and untapped resources. In those Neolithic times there were a million *Homo sapiens*; now there are 8 *billion* of us, soon to be 10 billion. Society has changed beyond all recognition, but one thing is the same: animals are bred and slaughtered for meat.

Factory farming is a cruel and wasteful process, shrouded in mythology; somehow seen as an 'efficient' way to produce protein, when it isn't. Animals reared in this way eat vast quantities of grain and waste most of its value in its conversion to meat. In this way, we waste crops enough to feed an extra 4 billion people.[100] Which is not to say that an extra 4 billion people on the planet all at once would be a good idea: it would be an environmental disaster. It is to underscore the wastefulness of the antiquated 'modernity' that is factory farming.

There's also the myth that the only way to get the 'right' protein is by eating meat and other products from livestock. The reality couldn't be more different: the majority of our protein already comes from plants. Meat provides just 18 per cent of our protein, dairy a tenth[101] – by current reckoning, about two-thirds of our protein comes from plant foods. Not so much from plant-based meat alternatives as everyday foods such as bread, beans, nuts and grains. However, years of false programming has left us unable to recognise it. I've lost count of how many times I've been introduced to a supermarket boss who's in charge of 'protein' only to find they are solely focused on meat. 'Protein means meat' has become hardwired in our psyche and rooted in our food culture – yet, it's wrong.

Maxed out

From those distant days of Mesopotamia, agriculture has expanded to occupy half the habitable land surface of the planet, and much of it is used to rear animals for human protein. Bundled together, livestock products – meat including fish, as well as dairy and eggs – contribute a total of 37 per cent of our protein and 18 per cent of calories worldwide. Yet, to produce what amounts to little more than a third of humanity's protein, we use 83 per cent of the world's farmland to rear livestock.[102] This is hardly efficient and leaves little room for growth.

In land-use terms, expanding livestock isn't really an option, unless we want to cut down the remaining forests and make the climate emergency even worse. As it is, climate scientists warn that our over-consumption of meat alone could trigger catastrophic climate change. It is a significant driver of a world that is warming up, of nature pushed to the edge, of declines in pollinating insects, of squandered antibiotics and of soils in a perilous state. And with most animal protein coming from industrial factory farms, it has become an urgent problem.

The way we produce protein clearly needs a rethink. To quote from Lampedusa's *The Leopard*, 'For things to remain the same, everything must change.'

It's time for a new protein paradigm, driven by a fusion of nature, technology and the business of food. We need to move farming away from its destructive industrial mode to a more creative, regenerative version, an approach where meat moves out of the Stone Age. As the former Saudi oil minister Sheikh Ahmed Zaki Yamani said, 'The Stone Age didn't end because we ran out of stones. It ended because we invented bronze tools, which were more productive.'[103]

If we didn't have them before, science has given us the means to end factory farming. While agribusiness interests defend their old ways of doing things, the case for change is becoming overwhelming – and is already happening.

Change is coming

Food companies are starting to realise that change is inevitable. Dedicated companies, led by Impossible Foods and Beyond Meat, have come out of nowhere to become icons of a new type of foods: plant foods that taste and feel just like meat. In addition, high street brands like KFC and Burger King have fallen over themselves to be part of the action. Tyson, the world's second biggest meat manufacturer, has invested $150 million in alternative proteins, including plant-based burgers that bleed. Backed by big investment, the technology involved has come on in leaps and bounds. No longer tasteless bits of cardboard, plant-based alternatives are giving 'real' meat a run for its money. As Pat Brown, CEO of Impossible Foods, told the beef industry media site Drovers, 'Unlike the cow, we get better at making meat every single day.'[104] Cargill, another of the world's largest meat manufacturers, has changed the name of its 'meat' department to 'protein', in recognition of the fact that the future of protein lies beyond meat alone. Billionaires like Bill Gates and Richard Branson have also invested in cultured or 'clean' meat grown from stem cells. Branson believes that thirty years from now, 'we will no longer need to kill any animals and that all meat will either be clean or plant-based, taste the same and also be much healthier for everyone'.[105]

As I write, more than seventy companies are developing cultured meat, which they expect to hit the shelves in waves. First will come burgers, sausages and nuggets grown from stem cells in bioreactors, replacing the ubiquitous ground meat that we currently get from factory-farmed cattle, chickens and pigs. Steaks will follow, replicating the structure and texture of tenderloin, T-bone or New York Strip. The next wave is likely to be cultured seafood, followed by 3D meat 'printers' that will sit next to your bread maker and deliver your favourite cuts at home.

Before all that arrives, affordable protein is likely to come from precision fermentation, a marriage of microbes and feedstock in fermenter tanks in the spirit of craft beer. The fundamental technology has been around since the 1980s, when it began to be used to produce human insulin. Before then, insulin for treating diabetes was expensive and had to be extracted from the pancreas of cows and pigs.[106] Market-leading alternative meat producer Quorn has been fermenting single-celled fungi for decades, but what's new is 'precision biology', the ability to programme microbes to produce specific products. With technological breakthroughs comes an acceleration in potential. Proteins from precision fermentation are predicted to match those from animal agriculture on price by 2025 and to cost five times less by 2030.[107]

As protein moves into a new era, the various technologies involved will feed off each other, making processes cheaper and even more efficient. Precision fermentation will be used to produce the specialised ingredients needed for plant-based alternatives and to make cheap and plentiful growth media for cultured meat.

Beyond limits

Whereas the production of protein from 'modern foods' is only just getting started, industrial animal agriculture has pretty much reached its limit in terms of scale and efficiency. According to the independent think tank RethinkX, new foods are likely to be ten times more efficient than a cow at converting feed into end product. And with less feed, less land is required to grow it, which means less water and waste.[108]

On the extremes of protein's new horizon, Air Protein in California is using techniques inspired by NASA to produce another generation of protein. In the run-up to the first lunar landings, scientists tried to learn how carbon dioxide exhaled by astronauts might be turned into food. The experiments were shelved, but Dr Lisa Dyson, CEO of Air Protein, has resurrected the idea, which she describes as being like making yogurt. In an interview with *Forbes*, she said, 'First, we start with elements from the air we breathe – carbon dioxide, oxygen and nitrogen – and combine these elements with water and mineral nutrients. Next, we use renewable energy and a probiotic production process where cultures convert the elements into nutrients. The output is a nutritious source of protein with the same amino acid profile as animal protein.'[109]

As each type of modern food becomes more efficient, so the inefficiencies within industrial animal agriculture will become harder to stomach. When the price of modern foods plummets, be they plant-based, cultured meats or made from precision fermentation or 'thin air', it will cut the legs off factory farming. Long hidden behind a veil of closed-door secrecy, misleading labelling and opaque government handouts, factory farming will come to be seen as the cruellest folly of our time. Like the slave trade, we will wonder how we let it happen.

Until then, industrial animal agriculture remains dominant in all its cruelty and crude inefficiency.

Factory farming's undoing might be the very thing that has propped it up against the odds for all this time: economics. Thanks to subsidies and a blind-eye approach to the hidden costs, the justification for factory farming has been its ability to produce 'cheap' food, but it is predicted that, by 2035, protein from modern food sources will be ten times cheaper than existing animal proteins. When this happens, it will cause major disruption. Factory farming became big business based on huge volumes and ultra-slim margins, but as the price dropped, so did the quality. Taste, nutritional value and ethics all went out the window. That low-profit margin and mass-production model has made factory farming vulnerable to the impact of a low-cost competitor. Why would anyone choose a factory meat nugget over a cheaper, tastier and healthier one produced from modern foods?

In the blink of an eye, the invincible industrial animal industry could tumble like a pack of cards. RethinkX predicts that the American cattle industry will start to collapse by 2030, with the number of cows falling by half and other industrial livestock sectors suffering 'a similar fate'. By 2035, it predicts that American demand for cow products will have fallen by up to 90 per cent. As product after product is undercut by modern foods, factory farming will go into a death spiral, with falling demand leading to higher prices and even lower demand. These may be predictions, but the direction of travel is clear.

If factory farming retreats, so will the croplands needed to feed factory-farmed animals. Half of the croplands of the EU, US and UK will no longer be needed for chemical-soaked monocultures of corn, soya and wheat to feed incarcerated animals. Great swathes of the countryside could be given back to nature, freed up for rewilding and reforestation. The farming countryside could be inherited by those visionary farmers who have shown that there is a better, kinder, more nature-friendly and regenerative way of producing food. The pasture-fed farmers who kept animals on the land in ways that brought back nature and increased soil fertility.

Those who encouraged bees, butterflies, birds and wildflowers. Those who stuck with it, despite decades of skewed economics favouring their cruel and destructive competitors.

Farmed animals will then be returned to the land, albeit in smaller numbers and on proper farms. They will play their part in rebuilding soil fertility, in mixed farms that create a diversity of products. Regenerative farmers will live beside the modern foods that contributed to the downfall of their industrial competitors. And perhaps, given time, regenerative farming will move beyond farmed animals altogether.

After 10,000 years of doing things the same way, humanity's love affair with steak and mince from cows could soon be over. As our food evolves, it will define all our futures, having a big bearing on the climate and nature emergency that threatens to pull us all down. But protein is stepping into a new dawn, and by 2100, meat-eating as we know it will be a thing of the past. Molecular gastronomy, powered by plants, microbes or cultured meat cells, will cause the demise of factory farming, and I for one will be happy to pronounce it dead.

17

REWILDING

BRINGING BACK WILD

Plunging dark clouds hung so heavy they barely seemed to clear the tops of the dense reeds that concealed four fledgling marsh harriers. For the briefest of moments, a breeze parted the stubborn stems to reveal a nest of male offspring, their brown heads flared with white caps. Cobweb wisps of downy white feathers showed that it would be a while before these naive creatures took to the air. The next day, 'ringers' were due to tag them with numbered metal rings around their legs and brightly coloured wing tags. When they finally take flight, eager watchers will document their every move. For now, they sit it out, waiting for their parents to return, getting fidgety when their big brown mum returns with a rodent, a toad or a hapless moorhen chick.

Over the years, I've spent many days watching marsh harriers, but I've rarely seen them as close as these ones at John and Charles Shropshire's Wissington Farm on the Fens, a stone's throw from where the dramatic bog oak discovery was made. These harriers were nesting beside the irrigation reservoir built by the Shropshire family to make sure their salad crops, onions, potatoes and sugar beets kept growing in the black, peaty soils during dry spells. As the presence of the harriers showed, they have not only built a successful business, they've also brought back wildlife. Nightingales, a nationally declining species, thrive here too, with twenty to thirty pairs gracing bushes with their enthralling song during spring. Like the marsh harriers, they are monitored; GPS trackers tell of their progress from the Fens to the Gambia or Senegal and back. The seventy miles of ditches on the farm also play host to returning water voles, perhaps best known as 'Ratty' in Kenneth Grahame's *Wind in the Willows*, another once-common species that is now on its uppers across much of the countryside.

The marsh harrier was the first big raptor that I 'found' for myself. As a young schoolboy during a family holiday, I was spooked by what seemed an impossibly large bird with broad, long wings rising up beside me. I became so enchanted by them that as a teenager I worked for several months as a volunteer at Titchwell Marsh RSPB.

Titchwell Marsh has since become a flagship nature reserve, a status well earned by the hard work put in by the veteran warden Norman Sills. I would hitchhike a hundred miles from Bedfordshire to north Norfolk and live in the cottage assigned to scruffy, young enthusiasts like me. I regret not making more of it, but I was too young and naive to learn as much as I should have. Back then, I wanted to follow in Norman's footsteps and become a nature reserve warden. But then I met a girl back in Bedfordshire, which rather took the shine off those distant days an eight-hour hitchhike away.

During my time at Titchwell, I remember with great fondness digging ditches, pulling bits of windblown hide out of the marsh, selling stuff in the shop and counting lots of birds. I helped make sure the little tern colony went undisturbed and watched over the first avocet chick to be hatched at Titchwell, a returning species that also features on the RSPB logo. Yet the thing I most remember was keeping a close eye on the marsh harriers as they flew over the reed beds, an icon of landscapes that to many might seem bleak.

What a surprise, forty years later, to learn that Norman had helped the Shropshires build the wetland on their farm, in order to attract marsh harriers and the like. From Titchwell, he had gone on to recreate the Lakenheath Fen, restoring it from carrot fields and in the process leading the return of cranes to the Fens. On his retirement, he told his local paper, 'When the Fens were the proper Fens, cranes nested there, but they didn't adapt to the new drained landscape and for the last 400 years there hasn't been anywhere in the Fens where they could come and nest. Lo and behold, in 2007 two pairs rolled up [at Lakenheath Fen] and the same two pairs are here today. You can get these nice surprises and it's part of the magic.'[1]

A decade later another milestone was reached, when the first crane chick in more than a century hatched at Wicken Fen, a National Trust reserve south-east of Ely and the Great Fen. In this heartland of British agriculture, it was encouraging to see wildlife being brought back from the brink and to see many farmers showing a real care for the countryside. While targeted conservation measures can bring back rare species like the crane, reversing declining wildlife more broadly requires changing the game.

Recognition is growing that more needs to be done. More farmers are searching for ways to escape the treadmill of intensification and become genuine stewards of the land, which means aiming to hand it over to the next generation in a better state than inherited. What remains against them is the way the cards have been stacked by decades of policies taking them in the wrong direction. Policies with serious consequences that a new generation of policymakers have one last chance to put right.

As we've seen, rapid human population growth and a shift to meat-based diets are increasing the pressure on existing farmland.[2] The typical response is to further intensify. To add yet more synthetic fertilisers and pesticides to monocultures of crops in an attempt to keep up with a treadmill that is gaining speed. The speed of that treadmill is ramped up several notches by feeding about a third of the world's harvest to factory-farmed animals. Throw in evidence that yields for many crops are plateauing[3] and the crisis starts to feel imminent. Yet more land gets cleared to make way for more crops, which means forests are raised and carbon released from rapidly depleting soils as the planet hurtles toward breaking point.

FARM WILDING

This collision of converging crises requires a plurality of solutions, in which nature-friendly farming has a big role to play. For too long, food production has ignored the fact that without care, finite resources like soils eventually run out. There is a real urgency about restoring farm animals to the land, but there is also a growing need to relieve the pressure on our precious land. Reducing the amount of meat and dairy produced would provide a quick win.

As a conservation-minded animal welfarist, I am passionate about the thought of rewilding, letting nature back in and restoring balance. I was keen to find out if there was a point where farming and rewilding meet, a way of providing for the needs of a growing population while not harming the planet of the future. To some, rewilding is a romantic but impractical notion. After all, if we're pressed for farmland already and the planet isn't making more land, how can we sacrifice fields to nature? Other people see it as the sustaining counterbalance to an agricultural machine that is driving wildlife off a cliff edge and bringing nature to a creaking halt. For rewilding to catch on beyond the fashionable whim of some wealthy landowners, it has to become part and parcel of how we think about farming.

One outstanding example of rewilding on agricultural land – and one that pays – is the Knepp Estate in West Sussex. Little more than half an hour from where I live, the 3,500-acre estate has since become the inspiration for many similar projects.

Over the years, the estate's owners, Charlie Burrell and Isabella Tree, have been kind enough to host me several times at Knepp and we see each other from time to time on the speaking circuit. To this day, I credit them with expanding my vision of what nature-friendly farming could look like. Their story is captured brilliantly in Isabella's book, *Wilding*.[4] I remember meeting Charlie when I was speaking at an event about the benefits of rearing ruminants purely on pasture. He came up to me and suggested that I ought to come and see how they were creating a woodland pasture on their farm. I'd become a bit of a sceptic towards such invitations but took up Charlie's nonetheless. I went to Knepp not knowing quite what to expect.

As I drove through the formal gates of the estate and past a perfectly choreographed deer park, my heart sank. My sense of disappointment deepened as I pulled up outside a crenellated castle. Surely this was a wasted trip? Charlie greeted me bleary-eyed, having just made an overnight trip from Scotland. I wasn't optimistic. We had a quick coffee, after which Charlie took me through the back door of his castle and out into the estate: it was an *Alice Through the Looking-Glass* moment, like stepping into another world.

Tired of trying to keep up with the intensive farming treadmill, Charlie had taken the decision some years earlier to just switch it off. To stop farming and let nature take over. Ten years on, a savannah of scrub, bushes, grasslands and coppices had replaced fields of crop monocultures. He then introduced animals that were close equivalents to the ancient species that shaped the European landscape centuries ago. So where there were once aurochs, tarpans, bison and wild boars, he brought in longhorn cattle, Exmoor ponies, red and fallow deer, and Tamworth pigs. Spotting them was tricky – they were essentially feral – but in the course of a morning, we located deer, ponies and cattle. It would take me three more trips to see the pigs. A Tamworth sow with hair as coarse as a wire brush had built a nest on the edge of woodland. I watched from a respectful distance as she sat amongst her energetic and curious piglets. Their eyes sparkled and their trotters raced. I couldn't help but see the puppy in them.

Knepp is widely celebrated for its returning wildlife – purple emperor butterflies, turtle doves, nightingales and other scarcities – as well as its range of 'free-roaming, pasture-fed, organic meats'.[5] Charlie is now working on bringing back other native wildlife species: white storks and beavers, to name just two. The estate pays for itself with income from seventy-five tonnes of meat a year, as well as nature-based tourism, glamping, property rentals and environmental stewardship payments. Since my first trip to Knepp, the world of solutions and possibilities has seemed so much bigger.

When I saw Charlie backstage at a Hampshire Wildlife Trust conference that we were both speaking at in 2019, I asked him where I should go to see the next farmland rewilding project. 'Who's the next Charlie Burrell?' 'Go to Holkham in Norfolk and talk to Jake Fiennes', he said.

NEW WAVE REVOLUTIONARY

England's Norfolk coast and the rain lashes down from sunken skies as Belted Galloway cattle graze on the pasture opposite. Sheep stand in the same field, separated by a dyke. Earlier, they were serenaded by thousands of pink-footed geese, a sound that defines the winter months along this coastline. I've been coming to Holkham National Nature Reserve since I was a boy. The coastline is a haven for wildlife, a magnet for naturalists, birders and wildlife buffs, but this time I'd come to see what was going on in the surrounding farmland.

Inside the smart new office of a converted barn, I was sitting with the most unlikely of farm revolutionaries. Jake Fiennes introduced himself as Charlie Burrell's longest-staying house guest: he went to Knepp for a weekend in the 1990s and ended up staying for four years. He and Charlie worked closely together on what was then an estate farmed intensively. Their love of wildlife blossomed at the same time as their mutual love of beer.

Forty-nine years old with piercing blue eyes, swept-back hair streaked white and dark and shaved at the sides, Fiennes had the air of a new wave rocker. I could see why he'd been described in the press as someone who could have taken up either farming or heroin.[6] I was picturing him sneering theatrically behind a microphone in gothic-punk garb. But in reality, he was dressed more conventionally for a farmer, cutting a distinctive figure in a blue shirt and smart green gilet.

The twin brother of the actor Joseph Fiennes, Jake was twelve months into the job as conservation manager for the Holkham Estate. His mission was to bolster wildlife across the 25,000-acre estate, from its intensively farmed arable land to its marshland habitats.

Holkham was one of the birthplaces of the agricultural revolution. During the early nineteenth century, the estate introduced the four-course 'Norfolk rotation', which was marked for using different crops to replenish soils rather than having a fallow period. It became standard practice on British farms and across much of continental Europe. The rotation typically had wheat followed by turnips, then barley, clover and pasture grazed by sheep.[7]

Sheep are still an important part of the rotation here at Holkham; Fiennes sees them as soil-fertility-building tools as well as lawnmowers for conservation grazing, keeping the grass the right length for wildlife. 'They're not fundamentally a meat product – the meat is a by-product of their role,' he told me.

Thomas William Coke, a politician and avid agriculturist, inherited the Holkham Estate in 1776 and built Holkham Hall, an elegant Palladian-style house that stands at the heart of the estate. Over the next forty years, 'Coke of Norfolk', as he became known, sparked an agricultural revolution that boosted food production and fostered sustainability. His annual three-day 'shearings' attracted farmers, aristocrats and royalty.[8]

Two hundred years later, Fiennes is out to inspire a new agricultural revolution. 'I'm trying to feed people in a way that isn't detrimental to nature,' he said. He sees the best way of doing that as by 'multifunctional farming' – feeding people through functioning ecosystems.

He has no time for the 'Taliban farming element' that has profited from intensive farming, with flailed hedges and crops sown right up to the edges of fields before being doused in chemicals. He thinks their days as numbered. He doesn't like twitchers – rare-bird enthusiasts – either, because, he says, they 'trample' areas where they shouldn't. I agreed with him about 'Taliban farming' but decided to keep my propensity for twitching well hidden. However, I did share with him that in 1984 I helped guard Britain's first nest of parrot crossbills, large finches with twisted beaks, in nearby woods, one of several volunteers who sat up all night at the base of the nesting tree to stop anyone shinning up and stealing the birds' eggs. When Fiennes later introduced me to his colleagues, I couldn't help noticing how my parrot crossbill vigil seemed to draw much more interest than anything else I'd done. Better that, I thought, than admit to being a twitcher.

For Fiennes, being accepted by the farming community was not straightforward. He had to get the language right. 'You can't speak to them about soils and worms and nature conservation –you have to speak to them in their language.' At his first National Farmers' Union conference, he was considered a square peg in a round hole. There wasn't much take-up for his breakout session on farming and the environment. Now that things have settled down, his sessions are oversubscribed and his plans for landscape-level change in farming are coming together. And, not surprisingly, his agricultural credentials are impressive, having a father who farmed in Suffolk and having himself worked on Charlie's Knepp Estate. Fiennes also spent a couple of years in Australia before managing Sir Nicholas Bacon's 5,000-acre Raveningham Estate for a decade and a half. A lot of his work, he believes, is about initiating a conversation about conservation. 'That's how you win hearts and minds.'

We hopped into his silver Ford Ranger pickup and drove off across the estate. We headed through the carefully manicured deer park around Holkham Hall, reminiscent of the one at Knepp. A group of deer bounded in front of the car just beyond the estate's dramatic stone arch entrance. As he smoked a roll-up, he took me past the lake where shoveler, gadwall and other ducks reside.

'Every spring and autumn on a farm, you have an opportunity to make changes', Fiennes told me, intent on using every one of those seasonal opportunities. At first seen as a 'disrupter', he has since got the backing of the farm director for the plans he wants to make across the estate's arable lands. To oil the wheels, he gave copies of Gabe Brown's *Dirt to Soil* to all the estate's leadership. Rotation is the key to this landscape, and he sees regenerative principles as important. 'We need to look after it [the soil] if it's to look after us.'

Fiennes is intent on moving Holkham away from growing maize; as I'm well aware from my experiences of the fields around home, it is renowned for being a 'needy' crop that leaves soils open to erosion. He believes in leaving corners and edges for nature and is not impressed with the way crops are still sown up to the field margins. To Fiennes, hedges are the 'backbone' of nature conservation, providing cover for wildlife and preventing soil erosion.

'I need to have a word with the man who cuts the hedges around here,' he said, trying not to show his annoyance. 'Why are we cutting hedges at all?' A fair question, I thought. Leave them to nature, and they'll provide a better windbreak and beetle bank for predatory insects and birds that feed on 'pests'. And as lifelong cattle farmer Rosamund Young points out in *The Secret Life of Cows*, for the welfare of farmed animals, 'the most important and versatile living shelter is a hedge.'[9]

More 'renaturing' than rewilding, I learned that blurring the lines of landscapes was part of his mission, together with letting nature recover on land that was uneconomic for farming. He told me that five hectares on the estate isn't seen as big enough for food production, so anything smaller is used for seed-bearing crops, hay meadows and other ways of boosting pollinators, birds and biodiversity.

Fiennes thinks having plenty of varieties of cover crops is important. 'Nature is diverse,' he said. 'The intersection of all these plant species benefits mycorrhizal fungi.' This is important in increasing the efficiency with which nutrients are captured for the next crop. The cover crops also provide tasty grazing for sheep and keep the soil covered during vulnerable times of the year. He sees the consequences of his neighbours not covering up. 'I see erosion everywhere. I see the roads full of silt. We need to prevent that.'

At Holkham, grazing animals are a big part of the system. Some six hundred adult cattle and two hundred calves graze the wet grasslands, which is crucial for birds and other wildlife. Fiennes showed me the new winter housing for the cattle, a Dutch barn with high open sides and plenty of straw. A colourful array of Belted Galloway, South Devonshire, Hereford and Limousin cattle bobbed their heads animatedly as they munched. They'd be out in the spring, 'mowing' the pastures; until then, they'll feed on species-rich grass silage cut from the surrounding area. In contrast to the acidic whiff of intensive silage, it smelt sweet. Four native grey partridges sheltered nearby under a dome-shaped hedge.

We drove on and reached Lady Anne's Drive, a long, straight road that runs to woodland and then dunes, with grazing marsh either side. This is the busiest car park on the north Norfolk coast, frequented each year by almost a million visitors, half a million cars and 300,000 dogs. A tonne of dog muck is removed from here each week.

The proximity of the birds was staggering. Wigeon, lapwing and gulls all stood close by. I voiced surprise, but Fiennes followed it with tales of grey partridge nesting within three feet of a parking space. 'I want to drown people in nature when they come,' he said.

In 2010, Holkham pulled off a real coup when spoonbills, tall white heron-like birds with long spatulate bills, nested for the first time in Britain. Since then, numbers have continued to increase. More than twenty pairs now breed here.[10]

Fiennes's vision is to get forty farmers together covering tens of thousands of hectares and really transform this land. He pictures a transformation based on respecting the soil, using 'catch' crops to literally grab nutrients and make them available to growing crops. He espouses regenerative farming using multiple species of plants and animals with much less chemical inputs.

'What's most important to a farmer? It's the soil beneath his feet. And I think we've kind of lost touch with that', he said. In many ways, he's keen to follow in the footsteps of Thomas William Coke and spark another agricultural revolution, based on a sympathetic and common-sense approach to farming. As he told me, it's 'not rocket science'. The good news for farmers is a growing amount of compelling evidence that wildlife-friendly farming practices, including creating nature habitats in marginal areas of the farm, can boost crop yields.[11]

Unlike Charlie Burrell, Fiennes doesn't see what he is doing as rewilding in the truest sense, more a practical approach where farming and wildlife coexist – producing food in welfare-friendly and nature-friendly ways today, while letting the soil and pollinators look after tomorrow. He sums up his vision in six words: 'Better meat and less of it.' In just a year at Holkham, he'd swept in a new culture and was only just getting started. So what's his motivation? 'It's for the good of my children's children,' he said.

REWILDING THE SOIL

Rewilding – the mass restoration of ecosystems – brings back nature and captures the imagination; surely, the next big opportunity lies in bringing back an 'elephant' of biodiversity to depleted farmland soils?

As a lifelong animal welfare environmentalist, the narrative during my lifetime has been all about farm animals being removed from the land, ecosystems being erased and species seeing their numbers decline. Rewilding has been a deep breath of fresh air, restoring nature and letting it take its course. Mountainsides and plains have been allowed to return to former glories. Rivers and valleys have been given the freedom to find their own way. New life has been created in previously lifeless uplands and charismatic species have returned home, from beavers in Britain to wolves in Yellowstone National Park.

Farms have also got in on the act – one of the best-known examples, of course, being the Knepp Estate, where the battle to make a living from difficult farmland has been won by letting nature run riot. Fields that have been intensively farmed have been given free rein to get back to nature, establishing a thriving woodland pasture where farm animals can live as wild among returning wildlife. It's an inspiring tale, as are many other rewilding projects.

Yet the biggest opportunity for rewilding relates to our depleted farmland soils, which present a massive canvas in desperate need of restoration. Soils are such rich habitats with so much biodiversity that they have been described as 'the poor person's rainforest'.[12] A healthy soil depends on a vibrant range of life forms below ground, from bacteria and fungi to insects, earthworms and moles. Soil organisms represent around a quarter of all biodiversity on Earth, yet they receive relatively little attention from conservationists.[13]

It's critical to understand that increasing the biodiversity of soil below ground relies on what happens above. As I have learned from pioneers like Will Harris, Gabe Brown, Charlie Burrell, Jake Fiennes and Jonny Rider, the secret to 'rewilding the soil' lies in keeping it covered, having a rich diversity of crops and moving free-ranging animals around the land in rotation. Grazing ruminants like cattle and sheep should be followed by clover-fed pigs or foraging chickens and maybe with ducks, turkeys and even goats added into the mix. Cover crops should be used to protect fields that would otherwise be bare when the main crop is over. Keeping the patchwork of different plants and animals moving round the farm, with a wide diversity of plants and animals in the mix, goes a long way to encouraging more biodiversity below ground.[14]

BRINGING BACK 'THE ELEPHANT'

But what about bringing back that elephant? Here, I'm not talking literally about elephants, but about the weight of life that should be in a single hectare of healthy soil. Treat it right and a hectare of arable land – little more than a football pitch – can hold as many as 13,000 species of life with a total weight of five tonnes – about the same as an elephant.[15]

To really bring soils to life, we need to feed that 'elephant' of biodiversity. Soil organisms thrive on dead organic matter, whether animal or plant remains, manure or residue. Underground organisms process it into nutrients that allow plants to flourish. They also protect plants from disease, help the soil store water, make nitrogen and other key elements more easily accessible and enable plants to communicate with each other. It is this richness of life that supports forests and prairies and enhances farming.[16]

So the secret to feeding the elephant within the soil is keeping it topped up with dead stuff and returning nutrients to the earth. Everything that lives on the land eventually dies. Microbes break it down, turning proteins, carbohydrates and lipids into the nitrogen, phosphorus and sulphur that plants need. By feeding the soil, plants are kept healthy naturally – and so is the soil.[17]

Soil is much more than just dirt – it is a living, breathing ecosystem. Up to 4 million worms can be found in a hectare of fertile land, potentially weighing more than the farmed animals above ground.[18] Soils without earthworms can be 90 per cent less effective at soaking up water.[19]

Soils also include some of the hardiest creatures on the planet. Tardigrades, also known as water bears, can be found in scorching deserts and frozen tundra and can even survive in space. These barrel-chested, eight-legged micro-animals digest food, making the nutrients available for plants.[20]

Underground fungi form vast underground networks that plants tap into via their roots and use to communicate with each other. Scientists have observed plants sending signals through fungal networks to warn neighbours of insect infestation, drought and other threats.[21] Then there are the 'litter transformers' that break stuff down into smaller bits, making it available to microbes – the micro-food web that includes nematodes and protozoa.[22]

All this richness of life is our 'silent ally', in the words of former UN FAO director general, José Graziano da Silva.[23] And then there is the economics: in 1997, the economic value of soil biodiversity was calculated at around $1.5 trillion per year,[24] and its stock is rising. While soil has declined since then, only now are we realising its value.

In the battle against climate change and the collapse of the natural world, rewilding the world's soils through animal 'welfare- and nature-friendly farming' offers a huge opportunity. We have the chance to integrate farmed animals and respect their well-being as sentient creatures, to provide good food naturally by dispensing with harmful chemicals and to add the 'elephant' of biodiversity that should be living beneath every step of farmland. Now that seems to me like a new frontier fit for the future.

EPILOGUE

NEW BEGINNINGS

Early morning, and distant trees were silhouetted against a reddening sky, while a veil of mist hung above the frost-brushed pasture. Birdsong was rising, with blackbirds, robins and song thrushes singing flutily in interweaving notes. A migrant whitethroat joined in from a ramshackle hedge of hawthorn, hazel and blackberry. On our front lawn, a male pheasant, resplendent with a long tail and fancy headdress, cooed softly with appreciation as he pecked at spilt seeds. Jackdaws jostled with mallards for leftovers. Later, they would be standing on the backs of cows, plucking beakfuls of hair to line their nests. The seasons had changed, and there was no time to lose.

Like the wildlife around our farm hamlet, I sensed change was in the air, and not just of the seasons: the whole world stood at a crossroads. It was April 2021 and Joe Biden, the new American president, had announced that humanity was entering a 'decisive decade' for tackling climate change.[1] In similar vein, the UK business secretary recognised that 'societal change', including diets with less meat, would help to meet climate targets.[2] And for my own small part, I had been chosen as one of a global network of 'champions' who would support the UN secretary general's call for 'transforming' food systems.[3] As I stood in the early-morning sun with my faithful canine companion Duke, I felt hopeful that the world might be waking up to the risks posed by climate change and the collapse of nature. There was still time to pull back from the brink.

***

So much of our societal thinking is based around the economy. Economists talk of 'corrections', sudden downturns in fortune for financial markets, and regard the Wall Street Crash of 1929 as a prime example. That crash led to the Great Depression and sparked the American Dust Bowl crisis, setting the scene for the emergence of factory farming. The sharp economic downturn caused by the coronavirus pandemic, leading to severe financial stress, is another, more recent example. Although hugely painful in the short term, such crashes and corrections are mercifully temporary, with economic growth eventually being restored. If we look at our planet in similar terms, we can see that we expect infinite growth in a finite world. And if we continue to exceed our planetary boundaries, nature's correction may be less forgiving. As it is, scientists describe the likely consequences of climate change as being 'so great that it is difficult to grasp for even well-informed experts'.[4]

While writing this book, I've discovered that change is inevitable, that if we carry on as we are, more planetary boundaries will be crossed. The prospect of climate change and the collapse of nature have entered the collective consciousness like an autumnal chill. While leaders are starting to *say* the right thing, only action will turn things around. Global agreements on climate change and the protection of biodiversity are already in place, but what's missing is a UN commitment to change the one thing that stands in the way of achieving success for both of these: food.

The question is, do we change the way we farm, or do we let it change us? For millennia, farming has worked in harmony with nature. But just one lifetime ago, things took a different course with the emergence of industrial agriculture. In reducing sentient creatures to the role of animal machines, we threw away our moral compass, disregarded the importance of farming *with* nature and tore up our 10,000-year-old contract with the soil. The massive black clouds of airborne soil that devastated the American plains during the Dust Bowl era were the early warning signs; pollution, the loss of pollinating insects and declining soils are the modern-day reminders. By the time we reach the hundredth anniversary of Black Sunday, the most notorious duster that hit Oklahoma in April 1935, we need to have made our food systems fit for the future. Key to this transformation will be how we address our meat- and dairy-rich diets that place such huge strain on the Earth's resources, with factory farming the driver of consumption.

When I look to the future, I am encouraged by the growth in farming methods that regenerate nature and by a willingness to change our diets. This offers the scope not only to preserve what we have, but to make it better – bring back wildlife, stabilise the climate and, for the first time in living memory, feed everyone in the world.

Regenerating the countryside relies on the restoration of farmed animals to the land as part of mixed, rotational farms, while balancing our diets with more ingredients from plants and alternative proteins. The joyous thing is that when animals are returned to the land in the right way – as rotational grazers or foragers – amazing things happen. Soils start to regenerate. As animals live more naturally, their droppings fertilise the land, harbouring insects that attract other wildlife and in turn become a source of nutrition. If we look hard enough, we'll find dung beetles taking parcels of manure into the soil, with worms, springtails and a host of other creatures starting to thrive. In every hectare of healthy soil, the mass of biodiversity equivalent to the weight of an elephant gets to work, cleaning roots, turning debris into compost and stimulating plant growth. Crop yields are restored, and animals are fed in ways that bring joy to their lives and respect their biology.

On regenerative and agroecological farms, the soil is held together by the mass of roots from pasture or cover crops. Renewed structure boosts crop growth, which makes soil more resilient to erosion and able to soak up vast quantities of rainwater, reducing the need for irrigation and preventing flooding.

Restoring farm animals to the land allows them to express their natural behaviours – running, flapping, grazing – making for happier animals with better immunity, which reduces the need for veterinary antibiotics. It cuts reliance on chemical pesticides and fertilisers, so reducing costs to farmers, and creates a varied landscape, bursting with wildflowers that lure back pollinating insects like bumblebees, as well as providing seeds and insects for birds and other wildlife. Rivers benefit from cleaner water, stronger banks and less flooding because the soil stays in the fields when it rains, while riverbeds no longer silt up, so trout and other fish have habitats in which to spawn. The water remains free of the carpets of green slime that come with polluting agricultural run-off. Nature-friendly, restorative farming rather than factory farming also takes the pressure off the world's remaining forests, reducing the need for deforestation to make way for more arable land. In this way, trees can carry on taking carbon out of the atmosphere in return for the oxygen we breathe.

Regenerative, nature-friendly farming is better able to feed the world; rather than being wastefully fed to animals, arable crops can nourish people directly, freeing up enough land to feed four billion people. By ending factory farming, therefore, we can feed everyone with less farmland, not more. And what a hugely healthy transition this could be. In the UK, where 55 per cent of cropland is used to grow animal feed, a third of that land could provide 62 million adults a year with their five daily portions of fruit and vegetables,[5] a demonstration of how transformative the move away from industrial agriculture could be. The 'Kray Twins' of food, factory-farmed chicken and bacon – culinary celebrities with dark secrets – would be consigned to the past, and we would no longer haul fish from the ocean to produce fishmeal for farmed chickens, pigs and fish. The world's seas could flourish, restoring penguins, sooty terns, seals, whales and sharks to their former glories, while the likelihood of new pandemics would recede along with the factory-farm pressure cooker that allows diseases to mutate and become more deadly.

Keeping animals on the land using regenerative farming practices helps to turbo-boost soil fertility and locks up carbon from the atmosphere – a modest increase in the carbon content of soil would offset the world's entire greenhouse gas emissions.[6] At the same time, reducing the number of farmed animals is essential if our food is to stay within planetary boundaries and avoid runaway climate change. The problem is that our appetite for meat is at an all-time high. Alternative proteins that hold the potential to transform our diets while satisfying our desire for meat could be our saving grace. Plant-based meat analogues, cultured meat and proteins from precision fermentation could replace meat and milk from animals with alternatives that look and feel authentic but don't have the downsides.

If we are to save the planet, by the middle of the century, we must reduce our consumption of livestock products globally by at least half, and high-consuming regions would need to make deeper cuts, with the UK and EU reining back on meat by two-thirds and the US by four-fifths. With time running out, alternative proteins are arriving at just the right time. They are the 'renewable energy' of the food sector, providing the power to charge our plates with meat, but without the downsides. Indistinguishable from animal meat, they offer the promise of being equally convenient to consumers and eventually at a lower price. Cultured meat and precision fermentation therefore offer the means to reduce our consumption of animal-sourced foods without taking anything away. Replacing one type of meat for another would be an unexpected easy win for policymakers. As it is, few governments yet have the stomach for drastic action on meat, but by the time planetary crisis kicks in, it could be too late. By then, actions would be about preventing things going from bad to worse. Forward-thinking governments and corporations have an opportunity to get behind alternative proteins through funding research and development and encouraging their uptake and, in doing so, could transform the food system.

There isn't just one solution that will bring about the necessary changes to our food system. We should embrace complexity and the range of solutions, including regenerative agroecological farming, plant-based diets, urban farming using hydroponics and aeroponics, alternative proteins from precision fermentation and cultured meat. All are essential ingredients on tomorrow's sustainable food menu. It will mean a veritable 'three Rs' approach – regenerative farming, reduction of animal-sourced foods and rewilding of the soil. Through a combination of regenerative farming and fewer farmed animals, the world's food system can become genuinely sustainable. Soil fertility can be turbo-boosted by that rotational symphony of plants and animals working in harmony with underground ecosystems. Huge amounts of carbon could be locked up in the ground. Meat from farmed animals would once again be an occasional treat, while everyday meat would come from cultured meat, plant-based proteins and other alternatives.

Although soil health in many parts of the world is pitiful and getting even worse, switching to regenerative farming would change things dramatically. And once soil health has been restored, society could reappraise the role of farmed animals. With meat from farmed animals, we will have reached a moment in history akin to what happened to the carthorse, where something that was once regarded as a necessity could become dispensable. Will the need for them be more about a conservation and land management technique than food? Will alternatives have made them redundant? The answers may vary around the world; Malawi, Nairobi or Delhi may answer differently to London, New York or Beijing. Whatever the answers, I am sure that meat-eating as we know it will be unrecognisable by the end of the century.

A departure from land-squandering industrial agriculture and an over-reliance on resource-hungry livestock will allow a mind-blowing resurgence of nature. Farming in harmony with nature – renaturing – combined with rewilding less productive land, holds the key to the restoration of wildlife. Habitats can be restored through the collective actions of neighbouring farmers, as the Burrells at Knepp and Jake Fiennes at Holkham have shown. Marginal lands and uplands could be reforested, restoring climate-stabilising trees to our hillsides and accelerating the return of wildlife, be it beavers, bison or mountain gorillas.

***

Back at home on that April day in 2021, I was reminded of the kinship that exists between living beings. I walked with Duke along the river valley in search of our neighbouring cows, a new herd freshly released from their winter housing. As we walked, we listened to the melodic rippling trill of a singing woodlark as a golden glow washed over the landscape.

After a while, we spotted the cows grazing on the far side of the river. Having noticed us, they gathered beside the bank and, one by one, waded across in a straggly line. Although we hadn't met these particular individuals before, I knew what would happen next. They were young, eager and inquisitive, bright-eyed and enthusiastic about life. The first ones out of the river fixed Duke with a stare. I could see his ears and nose twitching expectantly. They hesitated a bit before gingerly approaching. As they reached us, cows and dog extended their necks and sniffed each other curiously before going nose-to-nose. Then their tongues met in that gesture of kinship I remember seeing a year before – animals licking each other and exchanging saliva, as a means of greeting and reassurance. In that moment, I wanted the whole world to see what Duke was demonstrating: that these 'livestock', far from machines, were fellow creatures.

At the heart of sustainable change lies a recognition that all life on our planet is interconnected, and that our future depends on treating it with compassion and respect. In so doing, we can protect the world's wildlife and soils as if our life depends on it – because it does. The life expectancy of farmland soils would change from just sixty harvests left to one of infinite sustainability, while regenerative, agroecological farming would help end cruelty to animals, save wildlife, stabilise the climate and safeguard the planet for future generations. And to me, that seems like a future worth having.

ACKNOWLEDGEMENTS

In writing this book, I owe a huge debt of gratitude to everyone who has shaped the words and thinking within it. What started out as a one-off book, *Farmageddon*, has since turned into a trilogy – each book, and the thoughts within it, building on the next.

As with my previous two, this book has very much been a team effort. I am indebted to Jacky Turner for the many hours of painstaking research that form the backbone of *Sixty Harvests Left*. To Tina Clark, my long-suffering assistant: thank you for your hard work and encouragement. To Carol McKenna and Ali Large too, for endless support and input.

Huge appreciation to the trustees of Compassion in World Farming International for their sponsorship and support of the book; for their clear vision of how animal welfare and factory farming play a central role in driving planetary emergencies relating to climate, nature and health: Valerie James, Sir David Madden (whose literary mentorship I have appreciated hugely throughout), Sarah Petrini, Teddy Bourne, Professor Joy Carter, Joyce D'Silva, Jeremy Hayward, Mahi Klosterhalfen, Josphat Ngonyo and Reverend Professor Michael Reiss. And to former vice-chair Rosemary Marshall, to whose memory this book is dedicated.

Sincere thanks to Nick Humphrey for doing such a brilliant job of author's editorship as well as providing insight and advice throughout.

To my commissioning editor at Bloomsbury, Michael Fishwick, and his team: Amanda Waters, Elisabeth Denison, Lauren Whybrow, Kieron Connolly, Kathy Fry and Guy Tindale. Thanks also to my literary agent, Robin Jones.

To those who shaped the field trips where Covid allowed: Andrew Wasley and Luke Starr at Ecostorm; and Annamaria Pisapia, Federica di Leonardo and Mauricio Monteiro Filho, who undertook in Brazil what Covid prevented me from doing in person.

To those who shared important perspectives and help along the way, including Tim and Eddie Bailey, Sebastiano Cossia Castiglioni, Mike Clarke, Isha Datar of New Harvest, Henry Edmunds of Cholderton Estate, Duncan Grossart of The European Environment Trust, Graham Harvey, Seren Kell of the Good Food Institute, Professor Tim Lang, Leigh Marshall of Welney Wildfowl and Wetlands Trust, David Simmons of Riviera Produce, Rex Sly in the Fens, Professor Pete Smith, and Dr Tony Whitbread from Sussex Wildlife Trust.

Grateful thanks to all who commented on drafts and provided such helpful feedback, including Joyce D'Silva, Sean Gifford, Reineke Hameleers and Peter Stevenson. And for additional research and support: Jenny Andersson, Dr Krzysztof Wojtas and Emma Rush.

Finally, huge thanks to my wife, Helen, for her patience and understanding during long hours of drafting, and to Duke, my faithful companion on many walks of contemplation.

NOTES

PREFACE

1 Timothy Egan, *The Worst Hard Time*, First Mariner Books, New York, 2006, p. 204.

2 David Botti, Ashley Semler, Laura Trevelyan (producers), 'Filmmaker Ken Burns on the "dust bowl" drought', BBC News, 14 November 2012, https://www.bbc.co.uk/news/av/magazine-20301451/filmma ker-ken-burns-on-the-dust-bowl-drought; '1935 Black Sunday Dust Storm', On This Day in Weather History with Mark Mancuso, 21 April 2012, accuweather.com, https://www.youtube.com/watch?v=1OdD ieuD1OA; Ken Burns, *The Dust Bowl: Boise City Decline* PBS, aired 18 November 2012, https://www.pbs.org/video/dust-bowl-dust-bowl-boise-city-decline/?continuousplayautoplay= true.

3 Dayton Duncan and Ken Burns, *The Dust Bowl: An Illustrated History*, Chronicle Books: San Francisco, 2012, p. 95.

4 John Shaw, *This Land that I Love: Irving Berlin, Woody Guthrie, and the Story of Two American Anthems*, PublicAffairs, New York 2013.

5 Eduardo E. Zattara and Marcelo A. Aizen, 'Worldwide occurrence records suggest a global decline in bee species richness', *One Earth*, 22 January 2021, Vol. 4, Issue 1, pp. 114–23. https://www.cell.com/one-earth/pdfExtended/S2590-3322(20)30651-5.

6 R.E.A. Almond, M. Grooten and T. Petersen (eds.), *Living Planet Report 2020: Bending the curve of biodiversity loss* WWF, Gland, Switzerland, 2020, https://livingplanet.panda.org/en-us/.

7   Review on Antimicrobial Resistance, chaired by Jim O'Neill, Tackling drug-resistant infections globally: final report and recommendations, May 2016, https://amr-review.org/sites/default/files/160518_Final%20paper_with%20cover.pdf; Dame Sally Davies, UK Special Envoy on Antimicrobial Resistance, speaking at Healthy Food Systems: For people, planet, and prosperity, Independent dialogue for the UN Food Systems Summit, 4 June 2021.

8   Daniel Pauly and Dirk Zeller, 'Catch reconstructions reveal that global marine fisheries are higher than reported and declining', *Nature Communications*, 2021, 7:10244 DOI: 10.1038/ncomms10244 www.nature.com/naturecommunications; Steve Connor, 'Overfishing causing fish poplations to decline faster than thought, study finds', *Independent*, 19 January 2016, https://www.independent.co.uk/climate-change/news/overfishing-causing-fish-populations-to-decline-faster-than-thought-study-finds-a6821791.html.

9   Estimated from the global food waste and global livestock slaughtered per year. Sources are: Jenny Gustavsson, Christel Cederberg, Ulf Sonesson, Robert van Otterdijk and Alexandre Meybeck, 'Global food losses and food waste: Extent, causes and prevention', FAO, Rome, 2011, http://www.fao.org/3/mb060e/mb060e.pdf; Katie Flanagan, Kai Robertson and Craig Hanson, 'Reducing food loss and waste: setting a global action agenda', World Resources Institute, 2019, https://files.wri.org/s3fs-public/reducing-food-loss-waste-global-action-agenda_1.pdf; FAOSTAT database, 'Production, livestock primary', FAO, 2018, http://www.fao.org/faostat/en/#data/QL.

10 P. Pradhan et al., 'Embodied crop calories in animal products', *Environ. Res. Lett.*, 2013, 8, 044044, https://iopscience.iop.org/article/10.1088/1748-9326/8/4/044044/pdf; United Nations Convention to Combat Desertification (UNCCD), *Global Land Outlook*, UNCCD, Bonn, Germany, 2017, https://www.unccd.int/sites/default/files/documents/2017-09/GLO_Full_Report_low_res.pdf.

11 Emily S Cassidy, Paul C West, James S Gerber and Jonathan A Foley, 'Redefining agricultural yields: from tonnes to people nourished per hectare', *Environ. Res. Lett.* 8, 2013, 034015 doi:10.1088/1748-9326/8/3/034015.

12 FAO, 'Healthy soils are the basis for healthy food production', FAO, 2015, http://www.fao.org/3/a-i4405e.pdf.

13 Ibid.

14 FAO, 'World's most comprehensive map showing the amount of carbon stocks in the soil launched', FAO, Rome, 2017, http://www. fao.org/news/story/en/item/1071012/icode/.

15 P.R. Shukla et al. (eds.), *Climate Change and Land: An IPCC Special Report on climate change, desertification, land degradation, sustainable land management, food security and greenhouse gas fluxes in terrestrial ecosystems*, IPCC, 2019, https://www.ipcc.ch/srccl/.

16 Andy Challinor et al., *Climate and global crop production shocks*, Resilience Taskforce Sub Report, Annex A, Global Food Security Programme: UK, 2015, https://www.foodsecurity.ac.uk/publicati ons/resilience-taskforce-sub-report-annex-climate-global-crop-pro duction-shocks.pdf.

I BLACK GOLD

1 Rowan Mantell, 'Ancient tree buried in Norfolk field to be transformed into huge table', *Eastern Daily Press*, 5 August 2019, https://www. edp24.co.uk/news/ancient-fen-black-bog-oak-wissington-1-6198902.

2 'Fenland Black Oak: 5,000-year-old tree found in Norfolk', BBC News, 26 September 2012, https://www.bbc.co.uk/news/ uk-england-norfolk-19722595.

3 NFU East Anglia and NFU East Midlands, 'Delivering for Britain: Food and farming in the Fens', NFU, 2019, https://www. nfuonline.com/pcs-pdfs/food-farming-in-the-fens_web/.

4 NFU East Anglia and NFU East Midlands, 'Why farming matters in the Fens', NFU, 2008, https://www.nfuonline.com/assets/23991; FarmingUK Team 'New report explains why farming matters in the fens', *Farming UK*, 5 March 2008, https://www.farminguk.com/news/ new-report-explains-why-farming-matters-in-the-fens_6781.html.

5 Francis Pryor, *The Fens: Discovering England's Ancient Depths*, Head of Zeus, London, 2019.

6 'About the Great Fen: Heritage: Holme Fen Posts', Wildlife Trust for Bedfordshire, Cambridgeshire and Northamptonshire, https:// www.greatfen.org.uk/about-great-fen/heritage/holme-fen-posts.

7 The Rt Hon Michael Gove MP and Department for Environment, Food and Rural Affairs, 'The Unfrozen Moment: Delivering a Green Brexit' (Speech), gov.uk, 21 July 2017, https://www.gov.uk/governm ent/speeches/the-unfrozen-moment-delivering-a-green-brexit.

8 I.P. Holman, 'An estimate of peat reserves and loss in the East Anglian Fens', report commissioned by the RSPB, Cranfield University, October 2009, https://www.rspb.org.uk/globalassets/downloads/documents/positions/agriculture/reports/an-estimate-of-peat-reserves-and-loss-in-the-east-anglian-fens-.pdf.

9 Helena Horton, 'Farmers who help turtle doves should be rewarded with government cash, RSPB says', *Telegraph*, 17 November 2019, https://www.telegraph.co.uk/news/2019/11/17/farmers-help-turtle-doves-should-rewarded-government-cash-rspb/.

10 Fred Searle, 'G's joins forces with RSPB to save turtle dove', *Fresh Produce Journal*, 15 March 2017, http://www.fruitnet.com/fpj/article/171680/gs-joins-forces-with-rspb-to-save-turtle-dove.

11 Joe Pontin, 'Britain's tallest bird booms: common crane enjoys record year', *Countryfile Magazine*, 14 December 2018, https://www.countryfile.com/news/britains-tallest-bird-booms/.

12 Ian Newton, *Farming and Birds: Book 135*, Collins New Naturalist Library, London, 2017.

13 Hannah Ritchie and Max Roser, 'Crop Yields', *Our World in Data*, first published in 2017, substantial revision in September 2019, https://ourworldindata.org/crop-yields.

14 'How do you grow a record-breaking wheat crop? We spoke to the current and former record-holders to find out', Bayer Crop Science UK, https://cropscience.bayer.co.uk/blog/articles/2018/03/record-wheat-yield/#:~:text=Average%20wheat%20yields%20on%20UK,yield%20of%2016.52%20t%2Fha.

15 Ritchie and Roser, 'Crop Yields'.

16 D.B. Hayhow et al., *The State of Nature 2019*, State of Nature Partnership, 2019, https://nbn.org.uk/wp-content/uploads/2019/09/State-of-Nature-2019-UK-full-report.pdf.

17 Department for Environment, Food and Rural Affairs (DEFRA), 'Agricultural statistics and climate change', 9th edn., September 2019, https://assets.publishing.service.gov.uk/government/uploads/system/uploads/attachment_data/file/835762/agriclimate-9edition-02oct19.pdf.

18 Ian Rotherham, Sheffield Hallam University, personal communication, 22 November 2019.

19 NFU, 'Why farming matters in the Fens', FarmingUK Team, 'New report explains why farming matters in the fens'.

20 'Bardney Airfield History', Bomber County Aviation Resource website, http://www.bcar.org.uk/bardney-history.

21 NFU, 'Delivering for Britain'.

22 'NFU warns flood defence for Fens "inadequate"', BBC News, 17 May 2019, https://www.bbc.co.uk/news/uk-england-cambridgesh ire-48298361.

23 Challinor et al., *Climate and global crop production shocks*.

24 'About the Great Fen', Wildlife Trust for Bedfordshire, Cambridgeshire and Northamptonshire, https://www.greatfen.org.uk/about-great-fen.

25 Ibid.

26 Mark Ullyet, Reserves Officer, Great Fen project, Wildlife Trust for Bedfordshire, Cambridgeshire and Northamptonshire, personal communication, 25 November 2019.

2 A TALE OF TWO COWS

1 Rosamund Young, *The Secret Life of Cows*, Faber and Faber, London, 2017, p. 49.

2 Marcus Strom, 'Stand out from the herd: How cows commooonicate through their lives', University of Sydney, 19 December 2019, https://www.sydney.edu.au/news-opinion/news/2019/12/19/ stand-out-from-herd-how-cows-communicate.html.

3 Andrew Wasley and Heather Kroeker, 'Revealed: industrial-scale beef farming comes to the UK', *Guardian*, 29 May 2018, https:// www.theguardian.com/environment/2018/may/29/revealed-industr ial-scale-beef-farming-comes-to-the-uk.

4 'Brexit: Environmental and Animal Welfare Standards', UK Parliament debated on 20 July 2017, Hansard, Vol. 627, https:// hansard.parliament.uk/Commons/2017-07-20/debates/3087E 6DC-8EB6-4DFD-9B0E-AFE3C1968BB0/BrexitEnvironmentalAn dAnimalWelfareStandards.

5 R-Calf USA, 'Top 30 Cattle Feeders 2015', https://r-calfusa.com/ wp-content/uploads/2013/04/160125-Top-30-Cattle-Feeders.pdf.

6 Michelle Miller, 'Take a look inside one of the nation's largest cattle feedlots', AGDAILY, 2 July 2019, https://www.agdaily.com/livest ock/take-a-look-inside-one of the nations-largest-cattle-feedlots/.

7 Beef and Dairy Feeding / Locations, 'Choose between our Idaho or Washington feedlot location: Grand View, Idaho feedlot', Simplot website, http://www.simplot.com/beef_dairy_feeding/locations.

8   Michelle Miller, 'Take a look inside one of the nation's largest cattle feedlots'; Farm Babe, 'The Farm Babe™ unearths the truth behind modern farming', https://thefarmbabe.com/.

9   James S. Drouillard, 'Current situation and future trends for beef production in the United States of America: A review', *Asian-Australasian Journal of Animal Sciences*, July 2018, 31 (7), pp. 1007–16, https://www. ncbi.nlm.nih.gov/pmc/articles/PMC6039332/#:~:text=THE%20FEED LOT%20SECTOR,capacity%20greater%20than%201%2C000%20 animals.

10 Aisling Hussey, 'Karan Beef's feedlot in South Africa', *Irish Farmers Journal*, July 2015, https://www.youtube.com/watch?v=LAdB9O1Sd2A.

11 Felix Njini and Antony Sguazzin, 'China needs more beef. South Africa's trying to sell it', Bloomberg, 28 November 2018, https://www.bloomberg.com/news/articles/2018-11-28/ china-needs-more-beef-south-africa-s-trying-to-sell-it.

12 Jon Condon, 'Top 25 Lotfeeders: No 3 Whyalla Beef', Beef Central, 18 February 2015, https://www.beefcentral.com/features/top-25/lot-feeders/top-25-lotfeeders-no-3-whyalla-beef/.

13 Candyce Braithwaite, 'Exclusive: A tour of the largest feedlot in the southern hemisphere', *Sunshine Coast Daily*, 9 January 2017, https:// www.sunshinecoastdaily.com.au/news/take-a-tour-of-the-largest-feedlot-in-the-southern/3129755/.

14 Condon, 'Top 25 Lotfeeders'.

15 Braithwaite, 'Exclusive'.

16 FutureBeef / Knowledge Centre articles, 'Feedlots', 16 September 2011, reviewed 19 February 2018, https://futurebeef.com.au/knowle dge-centre/feedlots/.

17 Condon, 'Top 25 Lotfeeders'.

18 FutureBeef / Knowledge Centre articles, 'Feedlots'.

19 Lymington.com, 'All about pannage in the New Forest', https:// www.lymington.com/133-locally/1155-pannage-new-forest.

20 The New Forest / Explore, 'Wildlife and Nature', https://www.thene wforest.co.uk/explore/wildlife-and-nature.

21 Hampshire Ornithological Society, 'Hampshire Bird Sites: New Forest', https://www.hos.org.uk/home/news-recording/hampshire-birding/hampshire-bird-sites/new-forest/.

22 Royal Botanic Gardens Kew, 'Why Meadows Matter', 24 May 2017, https://www.kew.org/read-and-watch/meadows-matter.

23 Butterfly Conservation, 'The farmland butterfly and moth initiative', https://butterfly-conservation.org/our-work/england/the-farmland-butterfly-and-moth-initiative.

24 Hayhow et al., *The State of Nature 2019*.

25 British Trust for Ornithology: Birdtrends, 'Lapwing', https://app.bto.org/birdtrends/species.jsp?year=2019&s=lapwi.

26 Department for Environment, Food and Rural Affairs and Office for National Statistics, 'Wild Populations in the UK 1970 to 2019', DEFRA, 26 November 2020, https://assets.publishing.service.gov.uk/government/uploads/system/uploads/attachment_data/file/845012/UK_Wild_birds_1970-2018_final.pdf.

27 Ibid.

28 Ibid.

29 Ibid.

30 ShowMe England Online / Ross-on-Wye / Tourism, 'Symonds Yat Rock, Symonds Yat Villages', https://showmeengland.co.uk/ross-on-wye/tourism/symonds-yat-rock-symonds-yat-villages-king-arthurs-cave/; Wikishire / Wiki, 'Symonds Yat', https://wikishire.co.uk/wiki/Symonds_Yat.

31 'Greater Horseshoe Bat', School of Biological Sciences, University of Bristol, last modified 24 February 2005, http://www.bio.bris.ac.uk/research/bats/britishbats/batpages/greaterhorseshoe.htm#Status; J.S.P. Froidevaux et al., 'Factors driving population recovery of the greater horseshoe bat (*Rhinolophus ferrumequinum*) in the UK: implications for conservation', *Biodivers Conserv*, 2017, 26, pp. 1601–21 https://doi.org/10.1007/s10531-017-1320-1.

3 SIXTY HARVESTS LEFT

1 FAO, 'Where Food Begins', http://www.fao.org/resources/infographics/infographics-details/en/c/285853/.

2 Stephanie Pappas, 'Confirmed: The Soil Under Your Feet is Teeming with Life', Live Science, May 2016, https://www.livescience.com/54862-soil-teeming-with-life.html; Elaine R. Ingham, 'The living soil: Fungi', in *Soil Biology Primer*, University of Illinois, Urbana-Champaign, 2001, chapter 4, https://web.extension.illinois.edu/soil/SoilBiology/fungi.htm#:~:text=Along%20with%20bacteria%2C%20fungi%20are,and%20soil%20water%20holding%20capacity.

3 FAO, 'World's most comprehensive map showing the amount of carbon stocks in the soil launched', FAO, Rome, 5 December 2017, http://www.fao.org/news/story/en/item/1071012/icode/.

4 FAO, 'Healthy soils are the basis for healthy food production'.

5 FAO, 'World's most comprehensive map'.

6 Per Schjønning et al., 'Soil Compaction', in Jannes Stolte et al. (eds.), *Soil threats in Europe*, European Commission JRC Technical Reports, 2016, chapter 6, pp. 69–78, https://esdac.jrc.ec.europa.eu/public_ p ath/shared_folder/doc_pub/EUR27607.pdf.

7 David Geisseler and Kate M. Scow, 'Long-term effects of mineral fertilisers on soil microorganisms: A review', *Soil Biology and Biochemistry*, 2014, 75, 54e63, http://www-sf.ucdavis.edu/files/275520.pdf.

8 P.R. Shukla et al. (eds.), *Climate Change and Land: An IPCC Special Report on climate change, desertification, land degradation, sustainable land management, food security and greenhouse gas fluxes in terrestrial ecosystems*, IPCC, 2019, https://www.ipcc.ch/srccl/.

9 FAO, 'World's most comprehensive map'.

10 Ephraim Nkonya, Alisher Mirzabaev and Joachim von Braun, 'Economics of Land Degradation and Improvement: An Introduction and Overview', in *Economics of Land Degradation and Improvement: A Global Assessment for Sustainable Development*, Springer, Cham, Switzerland, 2016, Chapter 1, DOI 10.1007/978-3-319-19168-3_1.

11 Mary C. Scholes and Robert J. Scholes, 'Dust Unto Dust', *Science*, 2013, 342, pp. 565–6.

12 FAO, 'Nothing dirty here: FAO kicks off International Year of Soils 2015', FAO, Rome, 4 December 2014, http://www.fao.org/news/story/en/item/270812/icode/.

13 Arwyn Jones et al., *The State of Soil in Europe: A contribution of the JRC to the European Environment Agency's Environment State and Outlook Report – SOER 2010*, European Commission, 2012, http://publications.jrc.ec.europa.eu/repository/bitstream/JRC68418/lbna2 5186enn.pdf.

14 Francesco Morari, Panos Panagos and Francesca Bampa, 'Decline in organic matter in mineral soils', in Jannes Stolte et al. (eds.), *Soil threats in Europe*, European Commission JRC Technical Reports, 2016, Chapter 5, pp. 55–68, https://esdac.jrc.ec.europa.eu/public_ path/shared_folder/doc_pub/EUR27607.pdf.

15 Rothamsted Research, 'Careers at Rothamsted', https://www.rot hamsted.ac.uk/careers.

16 Rothamsted Research, 'About', https://www.rothamsted.ac.uk/about.

17 D. Allen and J. Crawford, 'Cloud structure on the dark side of Venus', *Nature*, 1984, 307, pp. 222–4, https://doi.org/10.1038/307222a0.

18 The Club of Rome, 'The Limits to Growth', https://www.clubofr ome.org/report/the-limits-to-growth/.

19 John Crawford, 'Global Agenda: What if soils run out?', World Economic Forum website, 14 December 2012, https://www.weforum. org/agenda/2012/12/what-if-soil-runs-out.

20 Chris Arsenault, 'Only 60 Years of Farming Left If Soil Degradation Continues', *Scientific American*, 5 December 2014, https://www.scientificamerican.com/article/only-60-years-of-farming-left-if-soil-degradation-continues/.

21 Duncan Cameron, Colin Osborne, Peter Horton FRS and Mark Sinclair, 'A sustainable model for intensive agriculture', University of Sheffield Grantham Centre for Sustainable Futures, 2015, http://grant ham.sheffield.ac.uk/wp-content/uploads/A4-sustainable-model-intensive-agriculture-spread.pdf.

22 H.K. Gibbs and J.M. Salmon, 'Mapping the world's degraded lands', *Applied Geography*, 2015, 57, pp. 12–21, https://reader.elsevier.com/rea der/sd/pii/S0143622814002793?token=6B9686092EBA0780DBF7B F902F838C6B2B3D92383E00022E90F3A5B26229C3C32E16E78EE C5C56AFB3DFB16B1A6AF46B.

23 R.J. Rickson et al., 'Input constraints to food production: the impact of soil degradation', *Food Security*, 2015, Vol. 7(2), pp. 351–64, https://link.springer.com/article/10.1007/s12571-015-0437-x; NationMaster: Arable land, 'Hectares: Countries Compared', https://www.nationmaster.com/country-info/stats/Agriculture/Arable-land/Hectares.

24 David R. Montgomery, *Dirt: The Erosion of Civilizations*, University of California Press, Berkeley, 2012, p. xii.

25 David Pimental and Michael Burgess, 'Soil Erosion Threatens Food Production', *Agriculture*, 2013, 3, pp 113–63, doi:10.3390/agriculture3030443.

26 USDA National Resources Conservation Service, 'Soil Formation: Washington Soil Atlas', https://www.nrcs.usda.gov/wps/

portal/nrcs/detail/wa/soils/?cid=nrcs144p2_036333#:~:text=An%20of
ten%20asked%20question%20is,%2C%20vegetation%2C%20
and%20other%20factors.

27 P.R. Shukla et al. (eds.), *Climate Change and Land: An IPCC Special
Report on climate change, desertification, land degradation, sustainable
land management, food security and greenhouse gas fluxes in terrestrial
ecosystems*, IPCC, 2019, https://www.ipcc.ch/srccl/.

28 Pimentel and Burgess, 'Soil Erosion Threatens Food Production'.

29 Grantham Centre for Sustainable Futures, 'Soil loss: an unfolding
global disaster: Grantham Centre briefing note', University of
Sheffield Grantham Centre for Sustainable Futures, 2 December
2015, http://grantham.sheffield.ac.uk/soil-loss-an-unfolding-global-
disaster/.

30 Crawford, 'Global Agenda'.

31 Grantham Centre for Sustainable Futures, 'Soil loss'.

4 THE MARCH OF THE MEGA-FARM

1 Stephen Burgen, 'Fears for environment in Spain as pigs
outnumber people', *Guardian*, 19 August 2018, https://www.
theguardian.com/world/2018/aug/19/fears-environment-spain-
pigs-outnumber-humans-pork-industry.

2 Ibid.

3 Food & Water Action Europe, 'Spain, Towards a pig factory farm
nation?', 15 March 2017, https://www.foodandwatereurope.org/repo
rts/spain-towards-a-pig-factory-farm-nation/.

4 Ibid.

5 'Foto-Natura-Huesca: Nature and Things of Interest', Miguel Angel
Bueno, the Vulture Shepherd website, https://foto-natura-huesca.
blogspot.com/2009/09/buitres-leonado-gyps-fulvus-gyps-fulvus.
html.

6 'Plans for Lincolnshire "super dairy" are withdrawn', BBC News, 16
February 2011, https://www.bbc.co.uk/news/uk-england-lincolnsh
ire-12485392.

7 Bristol City Council, 'The population of Bristol', https://www.bris
tol.gov.uk/statistics-census-information/the-population-of-bris
tol.

8 Claire Colley and Andrew Wasley, 'Industrial-sized pig and chicken
farming continuing to rise in UK', *Guardian*, 7 April 2020, https://

www.theguardian.com/environment/2020/apr/07/industrial-sized-pig-and-chicken-farming-continuing-to-rise-in-uk.

9 Compassion in World Farming, UK Factory Farming Map, 'Do you live in a factory farm hotspot?', https://www.ciwf.org.uk/fact ory-farm-map/#all/lincolnshire.

10 Andrew Wasley, Fiona Harvey, Madlen Davies and David Child, 'UK has nearly 800 livestock mega farms, investigation reveals', *Guardian*, 17 July 2017, https://www.theguardian.com/environm ent/2017/jul/17/uk-has-nearly-800-livestock-mega-farms-investigat ion-reveals.

11 Ecostorm, 2021. Research commissioned by Compassion in World Farming.

12 Mark Godfrey, 'China breeder increases slaughtering capacity', *FoodNavigator-Asia*, 12 March 2019). https://www.foodnaviga tor-asia.com/Article/2019/03/12/Muyuan-Foods-increases-slaughter-capacity#.

13 'Muyan Foodstuff Co To Boost Its Pig Herd', Large Scale Agriculture, 11 November 2019, https://www.largescaleagriculture.com/home/ news-details/muyan-foodstuff-co-to-boost-its-pig-herd/.

14 Agriculture and Horticulture Development Board, 'UK pig numbers and holdings', last updated 12 April 2021, https://ahdb.org.uk/pork/ uk-pig-numbers-and-holdings.

15 Compassion in World Farming Food Business: Award Winners, 'Muyuan Foodstuff Co. Ltd', https://www.compassioninfoodb usiness.com/award-winners/manufacturer/muyuan-foodstuff-co-ltd/.

16 Dominique Patton, 'Flush with cash, Chinese hog producer builds world's largest pig farm', Reuters, 7 December 2020, https://uk.rcut ers.com/article/us-china-swinefever-muyuanfoods/flush-with-cash-chinese-hog-producer-builds-worlds-largest-pig-farm-idUKKB N28H0CC.

17 Big Dutchman, 'Serving customers around the world: Housing and feeding equipment for modern pig production', 2/2013, https://cdn. bigdutchman.com/fileadmin/content/pig/products/en/Pig-product ion-Image-Big-Dutchman cn.pdf.

18 Big Dutchman: Pig Production, 'Breeding Stalls', https://www.big dutchmanusa.com/en/pig-production/products/sow-management/ breeding-stalls/.

19 Big Dutchman, 'Serving customers around the world', https://
    cdn.bigdutchman.com/fileadmin/content/egg-poultry/products/
    en/Egg-production-poultry-growing-Image-Big-Dutchman-en.
    pdf.

20 YouTube, 'UniVENT Conventional Layer Housing System', Big
    Dutchman North America: POULTRY, 8 January 2019, https://
    www.youtube.com/watch?v=lkwhpaHSGqA&feature=emb_logo.

21 The Rt Hon Michael Gove MP and Department for Environment,
    Food and Rural Affairs, 'The Unfrozen Moment: Delivering a
    Green Brexit', Speech, Gov.uk, 21 July 2017, https://www.gov.uk/
    government/speeches/the-unfrozen-moment-delivering-a-green-bre
    xit.

22 Council of the European Union, 'Outcome of Proceedings: EU
    priorities at the United Nations and the 75th United Nations General
    Assembly, September 2020–September 2021 – Council conclusions',
    Council of the European Union, Brussels, 13 July 2020, https://
    data.consilium.europa.eu/doc/document/ST-9401-2020-INIT/
    en/pdf?utm_source=POLITICO.EU&utm_campaign=8eb450c
    03e-EMAIL_CAMPAIGN_2020_07_15_04_59&utm_med
    ium=email&utm_term=0_10959edeb5-8eb450c03e-189131121.

23 United Nations Human Rights Office of the High Commissioner,
    '"Zero hunger" remains a distant reality for far too many, says UN
    expert', 5 March 2020, https://www.ohchr.org/EN/NewsEvents/
    Pages/DisplayNews.aspx?NewsID=25664&LangID=E.

5 CHAIN REACTION

1 YouTube, '*Correntão*', Laboratório Mecaniza – UFSM, 20 November
   2014, https://www.youtube.com/watch?v=wwzyK4-Qz3w;
   Wikipedia, 'Chains', https://translate.google.com/translate?hl
   =en&sl=pt&u=https://pt.wikipedia.org/wiki/Corrent%25C3%25
   A30&prev=search&pto=aue.

2 Trase, 'Sustainability in forest-risk supply chains: Spotlight on Brazilian
   soy', Trase Yearbook 2018, Transparency for Sustainable Economies,
   Stockholm Environment Institute and Global Canopy, https://yearb
   ook2018.trase.earth/ http://resources.trase.earth/documents/TraseY
   earbook2018.pdf; '2019 SoyStats: A reference guide to soybean facts
   and figures', American Soybean Association, 2019, https://soygrowers.
   com/wp-content/uploads/2019/10/Soy-Stats-2019_FNL-Web.pdf.

3 Ibid.

4 Walter Fraanje and Tara Garnett, 'Building Block – Soy: food, feed and land use change', Food Climate Research Network Foodsource, University of Oxford, 2020.

5 Tony Juniper, *Rainforest: Dispatches from Earth's most vital frontlines* Profile Books, London, 2018, pp. 134–6.

6 Ibid., pp. 164–6.

7 Ibid.

8 Dangerous Roads, 'Road BR-163: where trucks can get stuck for up to 10 days', https://www.dangerousroads.org/south-america/bra zil/2003-br-163.html.

9 Juniper, *Rainforest* pp. 164–6.

10 Trase, 'Sustainability in forest-risk supply chains'.

11 Associated Press, 'Brazil paves highway to soy production, sparking worries about Amazon destruction at a tipping point', MarketWatch, 12 December 2019, https://www.marketwatch.com/story/brazil-paves-highway-to-soy-production-sparking-worries-about-amazon-destruction-at-a-tipping-point-2019-12-12.

12 Ibid.

13 Liz Kimbrough, 'As 2020 Amazon fire season winds down, Brazil carbon emissions rise', Mongabay, 16 November 2020, https://news.mongabay.com/2020/11/as-2020-amazon-fire-season-winds-down-brazil-carbon-emissions-rise/.

14 Wikipedia, 'Munduruku', https://en.wikipedia.org/wiki/Mundur uku.

15 Federal Public Ministry, Attorney of the Republic in the Municipality of Santarém, 'Public Civil Action' (lawsuit to push back against agribusiness), Federal Public Ministry, Santarém, 29 May 2018, http://www.mpf.mp.br/pa/sala-deimprensa/documentos/2018/acao_ mpf_identificacao_delimitacao_territorio_munduruku_planalto_ santareno_pa_maio_2018.pdf .

16 Elizabeth Barona et al., 'The role of pasture and soybean in deforestation of the Brazilian Amazon', *Environ. Res. Lett.*, 2010, 5 024002, citing D. Nepstad, C.M. Stickler and O.T. Almeida, 'Globalization of the Amazon soy and beef industries: opportunities for conservation', *Conserv. Biol.*, 2006, 20 1595–1603, http://iopscie nce.iop.org/article/10.1088/1748-9326/5/2/024002?fromSearchP age=true.

17 E.Y. Arima et al., 'Statistical confirmation of indirect land use change in the Brazilian Amazon', *Environ. Res. Lett.*, 2011, 6 024010, http://iopscience.iop.org/article/10.1088/1748-9326/6/2/024010/meta.

18 Trase, 'Sustainability in forest-risk supply chains'.

19 Trase, 'The State of Forest Risk Supply Chains', Trase Yearbook 2020, Executive Summary, http://resources.trase.earth/documents/Trase_Yearbook_Executive_Summary_2_July_2020.pdf.

20 Daniel Nepstad and João Shimada, Earth Innovation Institute, 'Soybeans in the Brazilian Amazon and the Case of the Brazilian Soy Moratorium', International Bank for Reconstruction and Development / World Bank LEAVES Program, December 2018, Background Paper, https://www.profor.info/sites/profor.info/files/Soybeans%20Case%20Study_LEAVES_2018.pdf.

21 R.D. Garrett et al., 'Explaining the persistence of low income and environmentally degrading land uses in the Brazilian Amazon', *Ecology and Society*, 2017, 22, (3), p. 27, https://doi.org/10.5751/ES-09364-220327.

22 Associated Press, 'Brazil paves highway to soy production'

23 Reuters in Brasilia, 'Brazil's Amazon rainforest suffers worst fires in a decade', *Guardian*, 1 October 2020, https://www.theguardian.com/environment/2020/oct/01/brazil-amazon-rainforest-worst-fires-in-decade.

24 Ibid.

25 Stephen Eisenhammer, '"Day of Fire": Blazes ignite suspicion in Amazon town', Reuters, 11 September 2019, https://www.reuters.com/article/us-brazil-environment-wildfire-investiga-idUSKCN1VW1MK.

26 Ibid.

27 Dom Phillips and Daniel Camargos, 'Forest fire season is coming. How can we stop the Amazon burning?', *Guardian*, 5 May 2020, https://www.theguardian.com/environment/2020/may/05/a-deadly-cycle-of-destruction-how-greed-for-land-is-fuelling-amazon-fires.

28 Ibid.

29 Daniel Camargos and Dom Phillips, 'Fire Day "was the invention of the press", says principal investigated for burning in the Amazon', Repórter Brasil, 25 October 2019, https://reporterbrasil.org.br/2019/10/dia-do-fogo-foi-invencao-da-imprensa-diz-principal-investigado-por-queimadas-na-amazonia/.

30 Climate Action Tracker, 'Brazil – Country Summary', as at September
    2020, https://climateactiontracker.org/countries/brazil/.

31 Lucinda Elliott and Ben Webster, 'Deforestation rate in Amazon rises
    by a third', *The Times*, 19 November 2019, https://www.thetimes.co.uk/
    article/deforestation-rate-in-amazon-rises-by-a-third-whgloq92d.

32 Michael Krumholtz, 'Brazil's carbon emissions rising because of
    Amazon deforestation', Latin America Reports, 9 November 2020,
    https://latinamericareports.com/brazils-carbon-emissions-rising-
    because-of-amazon-deforestation/4809/.

33 Marianne Schmink et al., 'From contested to "green" frontiers in
    the Amazon? A long-term analysis of São Félix do Xingu, Brazil',
    Journal of Peasant Studies, 2019, 46:2, 377–99, https://www.tandfonl
    ine.com/doi/full/10.1080/03066150.2017.1381841; Gabriel Cardoso
    Carrero et al., 'Deforestation trajectories on a development frontier
    in the Brazilian Amazon: 35 years of settlement colonization, policy
    and economic shifts, and land accumulation', *Environmental
    Management*, 2020, 66, pp. 966–84, https://doi.org/10.1007/s00
    267-020-01354-w.

34 Ministério Público Federal, 'TRF1 annuls sentence that granted
    land reform area to farmers in Pará', 14 December 2016, http://
    www.mpf.mp.br/regiao1/sala-de-imprensa/noticias-r1/
    trf1-anula-sentenca-que-concedia-area-de-reforma-agraria-a-
    fazendeiros-no-para; http://www.ihu.unisinos.br/185-noticias/
    noticias-2016/557830-marcados-para-morrer-no-castelo-de-sonhos.

35 Umair Irfan, 'Brazil's Amazon rainforest destruction is at its highest
    rate in more than a decade', Vox, 18 November 2019, https://www.
    vox.com/science-and-health/2019/11/18/20970604/amazon-rainfor
    est-2019-brazil-burning-deforestation-bolsonaro.

6 A LAND WITHOUT ANIMALS

1 S. Hebron, 'An introduction to "To a Skylark"', British Library –
   Discovering Literature: Romantics and Victorians, https://www.bl.uk/
   romantics-and-victorians/articles/an-introduction-to-to-a-skylark.

2 Pierluigi Viaroli, University of Parma, personal communication, 17
   June 2016.

3 Giuseppe Zeppa, 'Grana Padano', Dairy Science Food Technology,
   2004, https://www.dairyscience.info/index.php/cheeses-of-the-
   piedmont-region-of-italy/82-grana-padano.html.

4 Cheese.com, 'Grana Padano', Worldnews, Inc., http://www.cheese. com/grana-padano/#.

5 Grana Padano, 'Specification of GRANA PADANO D.O.P'., http:// www.granapadano.it/assets/documenti/pdf/disciplinare_en.pdf.

6 Kees de Roest, *The Production of Parmigiano-Reggiano Cheese: The Force of an Artisanal System in an Industrialised World* Van Gorcum & Comp, Netherlands, 2000, p. 149, https://books.google.co.uk /books?id=VLTH05TqA8wC&pg=PA149&lpg=PA149&dq =italy+cheese+dairy+cow+zero+grazed&source=bl&ots =neuRkTeSvo&sig=mXjwqk_xcE_C3j3rl4qQvFUtyzY&hl =en&sa=X&ved=0ahUKEwiBzJf-wsvQAhVJLMAKHZRdC4EQ6 AEINDAE#v=onepage&q=italy%20cheese%20dairy%20cow%20z ero%20grazed&f=false.

7 Grana Padano, 'Specification of GRANA PADANO D.O.P.'

8 Cheese.com, 'Parmesan', Worldnews, Inc., http://www.cheese.com/ parmesan/.

9 Parmigiano Reggiano, 'Land', accessed June 2021, https://www. parmigianoreggiano.com/product-land/.

10 de Roest, *The Production of Parmigiano-Reggiano Cheese.*

11 FAOSTAT, Trade statistics, crops and livestock products.

12 YouTube, 'Quality Needs Compassion: The truth behind Italian hard cheeses', Compassion in World Farming, 8 March 2018, https:// www.youtube.com/watch?v=g0o6Cwlka30.

13 Council Directive 2008/120/EC, 'Laying down minimum standards for the protection of pigs', 18 December 2008, Annex I, Chapter I(8).

14 AnnoUno, Italian current affairs television programme (aired 21 May 2015).

15 UNESCO, 'Ferrara, City of Renaissance, and its Po Delta', UNESCO/NHK, http://whc.unesco.org/en/list/733.

16 B. Riedel, M. Zuschin and M. Stachowitsch, 'Dead zones: a future worst-case scenario for Northern Adriatic biodiversity', University of Vienna, 2008, http://homepage.univie.ac.at/martin.zuschin/ PDF/41_Riedel_et_al_2008c.pdf.

17 Austrian Science Fund, 'Without oxygen, "nothing goes" – Marine biologists get to the bottom of the dead zones', Phys.org, 26 July 2010, http://phys.org/news/2010-07-oxygen-marine-biologists- bottom.html.

18 Frank Ackerman, *Poisoned for Pennies: The Economics of Toxics and Precaution*, Island Press, Washington DC, 2008, https://books.google.co.uk/books?id=PYgmcdklVVUC&pg=PA105&lpg=PA105&dq=atrazine+when+banned+by+italy+%26+germany?&source=bl&ots=5vSa2lfSMZ&sig=dq-QntG8fbW7mlAZU8a5kZdZFzw&hl=en&sa=X&ved=0ahUKEwjGnfP2k87RAhWoDcAKHbHfAJIQ6AEINjAE#v=onepage&q=atrazine%20when%20banned%20by%20italy%20%26%20germany%3F&f=false; 'Science for Environment Policy – Herbicide levels in coastal waters drop after EU ban', European Commission DG Environment News Alert Service, edited by SCU, University of the West of England, 7 November 2013, Issue 349, http://ec.europa.eu/environment/integration/research/newsalert/pdf/349na2_en.pdf.

7 CLIMATE CRISIS

1 Brandon Specktor, '52 Polar Bears "Invade" a Russian Town to Eat Garbage Instead of Starve to Death', LiveScience.com, 2019, https://www.livescience.com/64741-polar-bears-are-taking-back-russia.html.

2 WWF, 'Experts will clarify the situation with polar bears on Novaya Zemlya Archipelago', WWF, Russia, 11 February 2019, https://wwf.ru/en/resources/news/arctic/wwf-ekspertam-predstoit-proyasnit-situatsiyu-s-belymi-medvedyami-na-novoy-zemle/.

3 NASA Global Climate Change, 'Facts: Arctic Sea Ice Minimum', https://climate.nasa.gov/vital-signs/arctic-sea-ice/.

4 National Wildlife Federation, 'Polar Bear', https://www.nwf.org/Educational-Resources/Wildlife-Guide/Mammals/Polar-Bear#:~:text=They%20mainly%20eat%20ringed%20seals,eat%20walruses%20and%20whale%20carcasses.

5 WWF, 'Experts will clarify the situation with polar bears.'

6 United Nations Secretary General Statement, *Secretary-General's statement on the IPCC Working Group 1 Report on the Physical Science Basis of the Sixth Assessment*, UN, New York, 9 August 2021, https://www.un.org/sg/en/content/secretary-generals-statement-the-ipcc-working-group-1-report-the-physical-science-basis-of-the-sixth-assessment.

7 IPCC, 'Summary for Policymakers of IPCC Special Report on Global Warming of 1.5°C approved by governments', IPCC, 2018, https://www.ipcc.ch/2018/10/08/summary-for-policymakers-of-ipcc-special-report-on-global-warming-of-1-5c-approved-by-governments/.

8 P.R. Shukla et al. (eds.), *Climate Change and Land: An IPCC Special Report on climate change, desertification, land degradation, sustainable land management, food security and greenhouse gas fluxes in terrestrial ecosystems*, IPCC, 2019, https://www.ipcc.ch/srccl/.

9 M. Springmann et al., 'Options for keeping the food system within environmental limits', *Nature*, 2018, 562, pp. 519–25, https://www.nature.com/articles/s41586-018-0594-0.

10 Ibid.

11 Ibid.

12 U. Skiba and Bob Rees, 'Nitrous oxide, climate change and agriculture', *CAB Review*, 2014, 9, 1–7, https://www.researchgate.net/publication/269403645_Nitrous_oxide_climate_change_and_agriculture.

13 H. Tian et al., 'A comprehensive quantification of global nitrous oxide sources and sinks', Nature, 2020, 586, pp. 248–56, https://doi.org/10.1038/s41586-020-2780-0.

14 United States Environmental Protection Agency, 'Overview of Greenhouse Gases: Nitrous Oxide Emissions', https://www.epa.gov/ghgemissions/overview-greenhouse-gases#nitrous-oxide.

15 C. D. Thomas et al., 'Extinction risk from climate change', *Nature*, 2004, 427(6970), pp. 145–8.

16 Ø. Wiig et al., '*Ursus maritimus*', IUCN Red List of Threatened Species, 2015, e.T22823A14871490, https://dx.doi.org/10.2305/IUCN.UK.2015-4.RLTS.T22823A14871490.en / https://www.iucnredlist.org/species/22823/14871490#population.

17 P.K. Molnár et al., 'Fasting season length sets temporal limits for global polar bear persistence', *Nature Climate Change*, 2020, 10, 732–8, https://doi.org/10.1038/s41558-020-0818-9.

18 Corey J. A. Bradshaw et al., 'Underestimating the Challenges of Avoiding a Ghastly Future', *Frontiers in Conservation Science*, 2021, Vol. 1, p. 9, https://www.frontiersin.org/article/10.3389/fcosc.2020.615419 / https://doi.org/10.3389/fcosc.2020.615419.

19 Richard Nield, 'Devastation and disease after deadly Malawi floods', Al Jazeera, 25 February 2015, https://www.aljazeera.com/features/2015/2/25/devastation-and-disease-after-deadly-malawi-floods.

20 'Malawi hit by armyworm outbreak, threatens maize crop', Reuters, 12 January 2017, https://www.reuters.com/article/us-malawi-grains-armyworms-idUSKBN14W0NT?feedType=RSS&feedName=environmentNews.

21 FAO, 'What is Soil Carbon Sequestration?', FAO, 2017. http://www.fao.org/soils-portal/soil-management/soil-carbon-sequestration/en/.

22 Rodale Institute, 'Regenerative organic agriculture and climate change', 2014, White Paper, http://rodaleinstitute.org/assets/WhitePaper.pdf.

23 Regenerative Organic Certified, 'Farm like the world depends on it', https://regenorganic.org/.

24 Quantis, 'General Mills: Accounting for soil impacts in carbon footprints, https://quantis-intl.com/casestudy/general-mills/.

25 Mariko Thorbecke and Jon Dettling, 'Carbon Footprint Evaluation of Regenerative Grazing at White Oak Pastures: Results Presentation', Quantis website, 25 February 2019, https://blog.whiteoakpastures.com/hubfs/WOP-LCA-Quantis-2019.pdf.

26 W.R. Teague et al., 'The role of ruminants in reducing agriculture's carbon footprint in North America', *J Soil and Water Conservation*, 2016, 71(2), pp. 156–64.

27 Communication from the Commission to the European Parliament, the European Council, the Council, the European Economic and Social Committee, the Committee of the Regions and the European Investment Bank, 'A Clean Planet for all: A European strategic long-term vision for a prosperous, modern, competitive and climate neutral economy', European Commission, Brussels, 28 November 2018, http://extwprlegs1.fao.org/docs/pdf/eur183103.pdf.

28 M. Springmann et al., 'Analysis and valuation of the health and climate change co-benefits of dietary change', *PNAS*, 2016, Vol. 113, No. 15, pp. 4146–51.

29 H. Harwatt, 'Including animal to plant protein shifts in climate change mitigation policy: a proposed three-step strategy', Climate Policy, 2018, DOI: 10.1080/14693062.2018.1528965;
H. Harwatt et al., 'Scientists call for renewed Paris pledges to transform agriculture', *Lancet Planet Health*, 2019 (published online 11 December), http://dx.doi.org/10.1016/S2542-5196(19)30245-1.

30 P. Smith et al., 'Agriculture, Forestry and Other Land Use (AFOLU)', in *Climate Change 2014: Mitigation of Climate Change, Working Group III Contribution to IPCC AR5*, Cambridge University Press, 2014.

31 Rob Bailey, Antony Froggatt and Laura Wellesley, 'Livestock – Climate Change's Forgotten Sector Global Public Opinion on Meat and Dairy Consumption', Chatham House, December 2014.

32 FAO, IFAD, UNICEF, WFP and WHO, *The State of Food Security and Nutrition in the World 2020:Transforming food systems for affordable healthy diets*, FAO, Rome, 2020, https://doi.org/10.4060/ca9692en.

33 TED, 'Greta Thunberg: The disarming case to act right now on climate change', TedxStockholm, November 2018, https://www.ted.com/talks/greta_thunberg_the_disarming_case_to_act_right_now_on_climate?language=en.

8 WILL INSECTS SAVE US?

1 L. Jackson, *East of England Bee Report: A report on the status of threatened bees in the region with recommendations for conservation action*, Buglife – Invertebrate Conservation Trust, Peterborough, 2019, https://www.wwf.org.uk/sites/default/files/201905/EofE%20bee%20report%202019%20FINAL_17MAY2019.pdf.

2 Yves Herman, 'Dutch firm generates buzz with big fly larvae farm', Reuters, 12 June 2019, https://www.reuters.com/article/uk-netherlands-insect-farm/dutch-firm-generates-buzz-with-big-fly-larvae-farm-idUKKCN1TC2O1?edition-redirect=uk; https://protix.eu/wp-content/uploads/Persbericht-Grand-Opening-ENG.pdf.

3 Julie J. Lesnik, 'Not just a fallback food: global patterns of insect consumption related to geography, not agriculture', Wiley Online Library, *American Journal of Human Biology*, first published 1 February 2017, https://doi.org/10.1002/ajhb.22976.

4 Arnold van Huis, *Edible Insects: Future Prospects for Food and Feed Security*, FAO, Rome, 2017, ISBN 9789251075968 OCLC 868923724.

5 Guiomar Melgar-Lalanne, Alan-Javier Hernández-Álvarez and Alejandro Salinas-Castro, 'Edible Insects Processing: Traditional and Innovative Technologies', Wiley Online Library, *Comprehensive Reviews in Food Science and Food Safety*, first published 30 June 2019, https://doi.org/10.1111/1541-4337.12463.

6 Ibid.

7 Katy Askew, 'Bugfoundation's vision: To change the eating habits of a whole continent', FoodNavigator.com, 12 October 2018, https://www.foodnavigator.com/Article/2018/10/12/Bugfoundation-s-vision-To-change-the-eating-habits-of-a-whole-continent#.

8 Jesse Erens et al., 'A Bug's Life: Large-scale insect rearing in relation to animal welfare', Wageningen UR students for MSc programme, September–October 2012, http://venik.nl/site/wp-content/uploads/2013/06/Rapport-Large-scale-insect-rearing-in-relation-to-animal-welfare.pdf.

9 M.E. Lundy and M.P. Parrella, 'Crickets are not a free lunch: protein capture from scalable organic side-streams via high-density populations of Acheta domesticus', *PLoS One*, 2015, 10(4), e0118785.

10 D.G. Oonincx et al., 'Feed conversion, survival and development, and composition of four insect species on diets composed of food by-products', *PLoS One*, 2015, 10(12), e0144601.

11 M. van der Spiegel, M.Y. Noordam and H.J. van der Fels-Klerx, 'Safety of Novel Protein Sources (Insects, Microalgae, Seaweed, Duckweed, and Rapeseed) and Legislative Aspects for Their Application in Food and Feed Production', Wiley Online Library, *Comprehensive Reviews in Food Science and Food Safety*, first published 15 October 2013, https://doi.org/10.1111/1541-4337.12032.

12 P. Brooke, 'Farming insects for food or feed', in *Farming, Food and Nature: Respecting Animals, People and the Environment* Routledge, Abingdon, 2018, p. 195.

13 M. Lechenet et al., 'Reducing pesticide use while preserving crop productivity and profitability on arable farms', *Nature Plants*, 2017, 3, 17008, https://www.nature.com/articles/nplants20178.

14 Brooke, 'Farming insects for food or feed', pp. 181–97.

15 C.A. Hallmann et al., 'More than 75 percent decline over 27 years in total flying insect biomass in protected areas', *PLoS One*, 2017, 12(10), e0185809, https://doi.org/10.1371/journal.pone.0185809.

16 Pedro Cardoso et al., 'Scientists' warning to humanity on insect extinctions', *Biological Conservation*, February 2020, Vol. 242, 108426 DOI: 10.1016/j.biocon.2020.108426.

17 Francisco Sánchez-Bayo and Kris A.G. Wyckhuys, 'Worldwide decline of the entomofauna: A review of its drivers', *Biological Conservation*, April 2019, Vol. 232, pp. 8–27.

9 WHEN OCEANS RUN DRY

1 YouTube, 'Escape Fishing with ET: Octopus Aquaculture', Australian Government Fisheries Research and Development Corporation, 23 December 2010, https://www.youtube.com/watch?v=9Iv2cBtgLCM.

2 Philip Hoare, 'Other Minds by Peter Godfrey-Smith review – the octopus as intelligent alien', *Guardian*, 15 March 2017, https://www.theguardian.com/books/2017/mar/15/other-minds-peter-godfrey-smith-review-octopus-philip-hoare.

3 National Geographic, 'Octopuses', https://www.nationalgeographic.com/animals/invertebrates/group/octopus-facts/#:~:text=Octopi%20have%20three%20hearts%20and%20blue%20blood.&text=Octopus%20skin%20is%20embedded%20with%20cells%20that%20sense%20light.

4 Gotowebinar, 'Pandemics, wildlife and intensive animal farming', CIWF EU webinar hosted by MEPs from all the key political groups and from seven different countries, 2 June 2020, https://register.gotowebinar.com/recording/8411919668736122886.

5 Jennifer Jacquet, Becca Franks, Peter Godfrey-Smith and Walter Sánchez-Suárez, 'The Case Against Octopus Farming', *Issues in Science and Technology*, Winter 2019, Vol. XXXV, No. 2, https://issues.org/the-case-against-octopus-farming/.

6 Hoare, 'Other Minds by Peter Godfrey-Smith review'.

7 C.F.E. Roper, M.J. Sweeney and C.E. Nauen, *FAO Species Catalogue: Cephalopods of the world, An annotated and illustrated catalogue of species of interest to fisheries – Octopuses (Order Octopoda)*, FAO Fish. Synop, 1984, No. 125, Vol. 3 http://www.fao.org/3/ac479e/ac479e32.pdf.

8 Warwick H.H. Sauer et al., 'World Octopus Fisheries', *Reviews in Fisheries Science & Aquaculture*, 2019, DOI: 10.1080/23308249.2019.1680603.

9 Jacquet, Franks, Godfrey-Smith and Sánchez-Suárez, 'The Case Against Octopus Farming.'

10 Ibid.

11 Ibid.

12 Rose Yeoman, 'Brave new world of octopus farming: Countering territorial behaviour and the propensity of octopus to escape from even the most securely closed tank systems have been among a number of achievements and world firsts to come from Australian efforts to develop aquaculture techniques for the species', *FISH*, Vol. 23, 1,

https://www.fishfiles.com.au/media/fish-magazine/FISH-Vol-23-1/ Brave-new-world-of-octopus-farming.

13 Ibid.

14 YouTube, 'Escape Fishing with ET: Octopus Aquaculture', Australian Government Fisheries Research and Development Corporation, 23 December 2010, https://www.youtube.com/watch?v=9Iv2cBtgLCM.

15 Group Nueva Pescanova, 'Researcher from Pescanova achieve to close the reproduction cycle of octopus in aquaculture', 18 July 2019, http://www.nuevapescanova.com/en/2019/07/18/researchers-from-pescanova-achieve-to-close-the-reproduction-cycle-of-octopus-in-aquaculture/#:~:text=%2D%20The%20Nueva%20Pescanova%20Group%2C%20headquartered,and%20companies%20around%20the%20world.

16 Ibid.

17 Jacquet, Franks, Godfrey-Smith and Sánchez-Suárez, 'The Case Against Octopus Farming'.

18 Fisheries Management Scotland, 'Fish Farming – North Carradale Escape', http://fms.scot/north-carradale-escape/.

19 Hamish Penman, 'Scotland's North Sea revenues take a hit as fiscal deficit increases to £15.1bn', Energy Voice, 26 August 2020, https:// www.energyvoice.com/oilandgas/north-sea/260923/north-sea-gers-figures-scotland/#:~:text=Sign%20Up-,Scotland's%20share%20 of%20North%20Sea%20oil%20and%20gas%20tax%20receipts,)%20 report%20for%202019%2D20.

20 The Fish Site, 'Salmon farming worth £2 billion to Scottish economy', The Fish Site, 29 April 2019, https://thefishsite.com/articles/salmon-farming-worth-2-billion-to-scottish-economy.

21 Scottish Salmon Producers Organisation, 'How much Scottish salmon is exported?', https://www.scottishsalmon.co.uk/facts/business/how-much-scottish-salmon-is-exported.

22 Aquaculture, 'Scotland, a land of food and drink: Aquaculture 2030', https://aquaculture.scot/.

23 'Scotland's wild salmon stocks "at lowest ever level"', BBC News Scotland, 24 April 2019, https://www.bbc.co.uk/news/uk-scotland-48030430.

24 'Charles launches "missing salmon" campaign', *Fish Farmer Magazine*, 28 November 2019, https://www.fishfarmermagazine.com/news/charles-launches-missing-salmon-campaign/.

25 Hannah Ritchie and Max Roser, 'Seafood Production', Our World in Data, 2019, https://ourworldindata.org/seafood-product ion#how-is-our-seafood-produced.

26 https://thefishsite.com/articles/a-new-high-for-global-aquaculture-production#:~:text=Total%20fish%20production%20is%20expec ted,26%20million%20tonnes)%20over%202018.

27 Compassion in World Farming calculation based on industry figures (per Dr K. Wojtas, 2021).

28 Feedback, *Fishy Business: The Scottish salmon industry's hidden appetite for wild fish and land*, Feedback, London, 2019, https:// www.feedbackglobal.org/wp-content/uploads/2019/06/ Fishy-business-the-Scottish-salmon-industrys-hidden-appet ite-for-wild-fish-and-land.pdf; Compassion in World Farming and One Kind, *Underwater cages, parasites and dead fish: Why a moratorium on Scottish salmon farming expansion is imperative*, CIWF & OneKind, March 2021, https://www.ciwf.org.uk/media/ 7444572/ciwf_rethink-salmon_21_lr_singles_web. pdf?utm_campaign=fish&utm_source=link&utm_medium=ciwf.

29 The Fish Site, 'A new high for global aquaculture production', The Fish Site, 8 June 2020, https://thefishsite.com/articles/the-preda tor-thats-killing-500-000-scottish-farmed-salmon-a-year.

30 Compassion in World Farming and Changing Markets Foundation, 'Until the Seas run dry: How industrial aquaculture is plundering the oceans', CIWF/Changing Markets, April 2019, https://www.ciwf. org.uk/media/7436097/until-the-seas-dry.pdf.

31 Ibid.

32 Louise Hunt, 'Fishmeal factories threaten food security in the Gambia', China Dialogue Ocean, 28 November 2019, https://chi nadialogueocean.net/11980-fishmeal-factories-threaten-food-secur ity-in-the-gambia/.

33 P. Veiga, M. Mendes and B. Lee-Harwood, 'Reduction fisheries: SFP fisheries sustainability overview', Sustainable Fisheries Partnership Foundation, 2018, https://www.sustainablefish.org/Media/Files/ Reduction-Fisheries-Reports/2018-Reduction-Fisheries-Report.

34 T. Baxter and P. Wenjing, 'China's distant water fishing industry is now the largest in West Africa', Unearthed, 24 November 2016, https://energydesk.greenpeace.org/2016/11/24/fishing-inside-chin ese-mega-industry-west-africa/.

35 Sanna Camara and Louise Hunt, 'Gambia's Migration Paradox: The Horror and Promise of the Back Way', New Humanitarian, 26 March 2018, https://www.newsdeeply.com/refugees/artic les/2018/03/26/gambias-migration-paradox-the-horror-and-prom ise-of-the-back-way.

36 Hannah Summers, 'Chinese fishmeal plants leave fishermen in the Gambia all at sea', *Guardian*, 20 March 2019, https://www.theguard ian.com/global-development/2019/mar/20/chinese-fishmeal-pla nts-leave-fishermen-gambia-all-at-sea.

37 Greenpeace, 'A waste of fish: Food security under threat from the fishmeal and fish oil industry in West Africa', Greenpeace International, Netherlands, June 2019, https://storage.googleapis.com/plan et4-international-stateless/2019/06/0bbe4b20-a-waste-of-fish-rep ort-en-low-res.pdf.

38 Compassion in World Farming and Changing Markets Foundation, 'Until the Seas run dry: How industrial aquaculture is plundering the oceans', CIWF/Changing Markets, April 2019, https://www.ciwf. org.uk/media/7436097/until-the-seas-dry.pdf.

39 FAO, *Code of Conduct for Responsible Fisheries*, FAO: Rome, 1995, http://www.fao.org/3/v9878e/V9878E.pdf.

40 Enrico Bachis, 'Fishmeal and fish oil: a summary of global trends', 57th IFFO Annual Conference, Washington, 2017, data for 2016.

41 Summers, 'Chinese fishmeal plants leave fishermen in the Gambia all at sea.'

42 Ibid.

43 Ibid.

44 Compassion in World Farming and Changing Markets Foundation, 'Until the Seas run dry'.

45 Christopher Feare, personal communication, 10 June 2018.

46 Norris McWhirter and Ross McWhirter, 'Loudest Pop Group', *Guinness Book of World Records*, Bantam Books, Toronto, 1973, p. 242, https://books.google.co.uk/books?id=Rv26phaJLUAC&q=Deep+Pur ple+loudest+intitle:Guinness&dq=Deep+Purple+loudest+inti tle:Guinness&redir_esc=y.

47 Feare, personal communication.

48 B. John Hughes et al., 'Long-term population trends of Sooty Terns *Onychoprion fuscatus*: implications for conservation status', *Popul*

Ecol, 2017, 59: 213–24, https://esj-journals.onlinelibrary.wiley.com/doi/epdf/10.1007/s10144-017-0588-z.

49 Christopher Feare, *Orange Omelettes and Dusky Wanderers: Studies and travels in Seychelles over four decades*, CPI Group, Croydon, 2016, p. 102.

50 C. Wilcox, E. Van Sebille and B.D. Hardesty, 'Threat of plastic pollution to seabirds is global, pervasive, and increasing', *Proc. Natl. Acad. Sci.*, 2015, 112, pp. 11899–904.

51 C.A. Erwin and B.C. Congdon, 'Day-to-day variation in sea-surface temperature reduces sooty tern sterna fuscata foraging success on the Great Barrier Reef, Australia', *Marine Ecology Progress Series*, 2007, 331, 255–66.

52 J.D. Reichel, 'Status and conservation of seabirds in the Mariana Islands', in J.P. Croxall (ed.), *Seabird status and conservation: a supplement*, International Council for Bird Preservation, Cambridge, 1991, pp. 249–62; Chris J. Feare, Sébastien Jaquemet and Matthieu Le Corre, 'An inventory of sooty terns (*Sterna fuscata*) in the western Indian Ocean with special reference to threats and trends', *Ostrich*, 2007, 78:2, 423–34, https://www.semanticscholar.org/paper/An-inventory-of-Sooty-Terns-(Sterna-fuscata)-in-the-Feare-Jaquemet/93c78b1c2fb093eaa7cd2b3a1eb0e2e3a60abc02.

53 Fisheries Global Information System (FIGIS), UN FAO, Rome, 1950-2019.

54 European Commission, Oceans and fisheries, 'Seychelles: Sustainable fisheries partnership with Seychelles', accessed June 2021, https://ec.europa.eu/oceans-and-fisheries/fisheries/international-agreements/sustainable-fisheries-partnership-agreements-sfpas/seychelles_en.

55 S. Cramp et al., *Handbook of the Birds of Europe, the Middle East and North Africa: The Birds of the Western Palearctic*, Oxford University Press, 1985, Vol. IV, p. 117.

56 Ibid.

57 Feare, *Orange Omelettes and Dusky Wanderers*, pp. 315–16.

58 Karen Green, 'Fishmeal and fish oil facts and figures', Seafish, March 2018, https://www.seafish.org/media/Publications/Seafish_FishmealandFishOil_FactsandFigures2018.pdf.

59 D. Gremillet et al., 'Persisting Worldwide Seabird-Fishery Competition Despite Seabird Community Decline', *Current Biology*, 17 December 2018, Vol. 28, Issue 24, pp. 4009–13, https://www.sciencedirect.com/science/article/pii/S0960982218314180.

60 S. James Reynolds et al., 'Long-term dietary shift and population decline of a pelagic seabird – A health check on the tropical Atlantic?', Wiley Online Library, *Global Change Biology*, 2019, https://doi.org/10.1111/gcb.14560.

61 Josh Gabbatiss, 'Seabirds on British island decline by 80% after overfishing and climate change cut off food source', *Independent*, 4 February 2019, https://www.independent.co.uk/environment/seabirds-colony-fishing-climate-change-ascension-island-atlantic-ocean-a8758941.html.

62 J. Del Hoyo, A. Elliott and J. Sargatal (eds.), *Handbook of the Birds of the World*: *V3 Hoatzin to Auks*, Lynx Edicions, Barcelona, 1996, Vol. 3, pp. 642–3.

63 D. Gremillet et al., 'Persisting Worldwide Seabird-Fishery Competition.'

64 D. Gremillet et al., 'Starving seabirds: unprofitable foraging and its fitness consequences in Cape gannets competing with fisheries in the Benguela upwelling ecosystem', *Marine Biology*, 2016, 163, 35, https://journals.plos.org/plosone/article/file?id=10.1371/journal.pone.0210328&type=printable.

65 Kenneth Brower, 'Life in Antarctica Relies on Shrinking Supply of Krill', *National Geographic*, 17 August 2013, https://www.nationalgeographic.com/animals/article/130817-antarctica-krill-whales-ecology-climate-science.

66 Whale Facts, 'What Do Blue Whales Eat? Diet, Eating Habits and Consumption', https://www.whalefacts.org/what-do-blue-whales-eat/.

67 Commission for the Conservation of Antarctic Marine Living Resources, 'Krill fisheries and sustainability', https://www.ccamlr.org/en/fisheries/krill-fisheries-and-sustainability.

68 Commission for the Conservation of Antarctic Marine Living Resources, 'Krill fisheries', https://www.ccamlr.org/en/fisheries/krill.

69 A. Atkinson, V. Siegel, E. Pakhomov et al., 'Long-term decline in krill stock and increase in salps within the Southern Ocean', *Nature*, 2004, 432, 100–3, https://doi.org/10.1038/nature02996. https://www.nature.com/articles/nature02996.

70 Fisheries Global Information System (FIGIS) Database, 'Capture statistics', UN FAO, Rome 1950-2019.

71 Aker BioMarine and Qrill Aqua, Contact Us, https://www.qrillaqua.com/contact-qrill-aqua.

72 Rimfrost, Contact Rimfrost, https://www.rimfrostkrill.com/contact.

73 Josh Gabbatiss, 'Krill fishing industry backs massive Antarctic ocean sanctuary to protect penguins, seals and whales', *Independent*, 9 July 2018, https://www.independent.co.uk/environment/antarctica-krill-fishing-industry-marine-protected-zone-greenpeace-whales-seals-penguins-a8439311.html.

74 Ibid.

75 Andrea Kavanagh, 'Off Antarctic Peninsula Concentrated Industrial Fishing for Krill is Affecting Penguins', Pew, 20 February 2020, https://www.pewtrusts.org/en/research-and-analysis/articles/2020/02/20/off-antarctic-peninsula-concentrated-industrial-fishing-for-krill-is-affecting-penguins.

76 Matthew Taylor, 'Decline in krill threatens Antarctic wildlife, from whales to penguins', *Guardian*, 14 February 2018, https://www.theguardian.com/environment/2018/feb/14/decline-in-krill-threatens-antarctic-wildlife-from-whales-to-penguins.

77 L. Hückstädt, 'Crabeater Seal – *Lobodon carcinophaga*', IUCN Red List of Threatened Species, 2015, e.T12246A45226918, https://dx.doi.org/10.2305/IUCN.UK.2015-4.RLTS.T12246A45226918.en.

78 G.J.G Hofmeyr, 'Antarctic Fur Seal – *Arctocephalus gazella*', IUCN Red List of Threatened Species, 2016, e.T2058A66993062, https://dx.doi.org/10.2305/IUCN.UK.2016-1.RLTS.T2058A66993062.en.

79 Del Hoyo, Elliott and Sargatal (eds.), *Handbook of the Birds of the World*, Vol. 1, pp. 140–53.

80 Mukhisa Kituyi, Secretary General, United Nations Conference on Trade and Development (UNCTAD) and Peter Thomson, United Nations Secretary General's Special Envoy for the Ocean, United Nations, '90% of fish stocks are used up – fisheries subsidies must stop emptying the ocean', World Economic Forum, 13 July 2018, https://www.weforum.org/agenda/2018/07/fish-stocks-are-used-up-fisheries-subsidies-must-stop/.

81 FAO, 'State of the world fisheries and aquaculture', *SOFIA*, 2018.

10 THE DUST BOWL ERA

1 Kevin Baker, '21st Century Limited', from *Harper's Magazine*, in Andrew McCarthy (ed.), *The Best American Travel Writing 2015*, HMHCo, New York, 2015, p. 51, https://books.google.co.uk/books?id=fvKOCgAAQBAJ&pg=PA51&lpg=PA51&dq=dust+bow

l+Riches+in+the+soil,+prosperity+in+the+air,+progress+everywh
ere&source=bl&ots=W9s4c4nepo&sig=ACfU3U38_dTooyZ
ZABNKqLoZ_cZX_LP6AQ&hl=en&sa=X&ved=2ahUKEwi_
i77wldnkAhWkQUEAHWRCBfUQ6AEwCnoECAcQAQ#v=onep
age&q=dust%20bowl%20Riches%20in%20the%20soil%2C%20
prosperity%20in%20the%20air%2C%20progress%20everywh
ere&f=false.

2 *Colorado Experience: The Dust Bowl*, Rocky Mountain PBS, 10 October
2014, https://www.youtube.com/watch?v=RKSvqTzgMrA.

3 The Wild West Pioneer, 'Homestead Act', https://sites.google.com/a/
comsewogue.k12.ny.us/the-wild-west-pioneer/leading-stories/
homestead-act.

4 'Thomas Jefferson's Presidency: What was Thomas Jefferson's vision
for the United States?', enotes, https://www.enotes.com/homew
ork-help/what-was-thomas-jeffersons-vision-united-states-277
769.

5 Caroline Henderson, *Letters from the Dust Bowl*, edited by Alvin
O. Turner, University of Oklahoma Press, 2003, p. 33.

6 History, 'Dust Bowl', 27 October 2009, updated 5 August 2020,
https://www.history.com/topics/great-depression/dust-bowl.

7 Dayton Duncan and Ken Burns, *The Dust Bowl: An Illustrated
History*, Chronicle Books, San Francisco, 2012, p. 25.

8 Timothy Egan, *The Worst Hard Time*, First Mariner Books, New York,
2006, p. 19.

9 Donald Worster, *Dust Bowl: The Southern Plains in the 1930s*, Oxford
University Press, 2004, https://books.google.co.uk/books?id=8fM-
ZWXPe_QC&pg=PA244&lpg=PA244&dq=donald+worster+natu
re%27s+winning+design&source=bl&ots=thof_Ny871&sig=ACfU
3U35kVnkxA3BOx8yjNis8kA-8bD9Hw&hl=en&sa=X&ved=2ah
UKEwjawPaQgqHlAhXNVsAKHVvoDGsQ6AEwEnoECAYQA
Q#v=onepage&q=donald%20worster%20nature's%20winning%20
design&f=false.

10 Egan, *The Worst Hard Time*.

11 Legends of America, 'Buffalo Hunters', https://www.legendsofamer
ica.com/we-buffalohunters/.

12 The Editors of Encyclopaedia Britannica, 'Dust Bowl', *Encyclopedia
Britannica*, 13 March 2021, https://www.britannica.com/place/
Dust-Bowl.

13 'Great American Desert', Colorado Encyclopedia, adapted from Martyn J. Bowden, 'Great American Desert', in David J. Wishart (ed.), *Encyclopedia of the Great Plains*, University of Nebraska Press, Lincoln, 2004, https://coloradoencyclopedia.org/arti cle/%E2%80%9Cgreat-american-desert%E2%80%9D; Digital History, 'The Great American Desert', http://www.digitalhistory. uh.edu/disp_textbook.cfm?smtid=2&psid=3148.

14 Egan, *The Worst Hard Time*, p. 50.

15 Henderson, 'Letters from the Dust Bowl', *Atlantic Monthly*, May 1936, p. 151.

16 Donald Worster, *Dust Bowl*, Handbook of Texas Online, Texas State Historical Association, https://tshaonline.org/handbook/online/ articles/ydd01.

17 Egan, *The Worst Hard Time*, p. 53; Duncan and Burns, *The Dust Bowl*, p. 29.

18 Historic Events for Students: The Great Depression, 'Dust Bowl 1931–1939', Encyclopedia.com, 15 April 2021, https://www. encyclopedia.com/education/news-and-education-magazines/ dust-bowl-1931-1939.

19 Duncan and Burns, *The Dust Bowl*, p. 37.

20 Egan, *The Worst Hard Time*, p. 59.

21 Historic Events for Students, 'Dust Bowl 1931–1939'.

22 Paul D. Travis and Jeffrey B. Robb, 'Wheat', Encyclopedia of Oklahoma History and Culture, https://www.okhistory.org/publicati ons/enc/entry.php?entry=WH001.

23 *The Great Depression Hits Farms and Cities in the 1930s*, Iowa PBS, http://www.iptv.org/iowapathways/mypath.cfm?ounid=ob_000064.

24 *The Dust Bowl, A Film by Ken Burns*, https://www.pbs.org/kenburns/ dustbowl/legacy/.

25 Egan, *The Worst Hard Time*, pp. 113–14; Duncan and Burns, *The Dust Bowl*, p. 42.

26 Bill Ganzel, Ganzel Group, 'Farming in the 1930s: The Dust Bowl', Wessels Living History Farm – York, Nebraska, first written and published in 2003, https://livinghistoryfarm.org/farminginthe30s/water_02.html.

27 Historic Events for Students, 'Dust Bowl 1931–1939'.

28 Duncan and Burns, *The Dust Bowl*, p. 43.

29 Henderson, *Letters from the Dust Bowl* ('Dust to Eat', letter to Henry A. Wallace, Secretary of State, 26 July 1935), pp. 140–7.

30 Historic Events for Students, 'Dust Bowl 1931–1939'.

31 Duncan and Burns, *The Dust Bowl*, p. 78.

32 Egan, *The Worst Hard Time*, p. 235.

33 History, 'Dust Bowl'.

34 Duncan and Burns, *The Dust Bowl*, pp. 54–6.

35 Egan, *The Worst Hard Time*, p. 141; Historic Events for Students, 'Dust Bowl 1931–1939'.

36 Duncan and Burns, *The Dust Bowl*, pp. 56–7.

37 Henderson, *Letters from the Dust Bowl* ('Spring in the Dust Bowl', *Atlantic Monthly*, 1937), p. 164.

38 Ibid., p.107.

39 Franklin D. Roosevelt, Inaugural Address, 4 March 1933, as published in Samuel Rosenman (ed.), *The Public Papers of Franklin D. Roosevelt* (Random House: New York, 1938), Vol. 2, pp. 11–16, audio extract of speech at: http://historymatters.gmu.edu/d/5057/.

40 National Park Service, 'FDR's Conservation Legacy', accessed June 2021, https://www.nps.gov/articles/fdr-s-conservation-legacy.htm.

41 Paul M. Sparrow, FDR Library, 'FDR and the Dust Bowl', By Fdrlibrary, Posted In From The Museum, 20 June 2018, https://fdr.blogs.archives.gov/2018/06/20/fdr-and-the-dust-bowl/.

42 Henderson, *Letters from the Dust Bowl*.

43 Duncan and Burns, *The Dust Bowl*, p. 68.

44 R. Douglas Hurt, *Documents of the Dust Bowl*, ABC-CLIO: Santa Barbara, California, 2019.

45 USDA National Resources Conservation Service, 'Biography of Hugh Hammond Bennett, April 15, 1881 — July 7, 1960, The Father of Soil Conservation', https://www.nrcs.usda.gov/wps/portal/nrcs/detail/national/about/history/?cid=nrcs143_021410.

46 Egan, *The Worst Hard Time*, p. 267.

47 Hannah Holleman, *Dust Bowls of Empire: Imperialism, Environmental Politics, and the Injustice of 'Green' Capitalism*, Yale University Press, New Haven, Connecticut, 2018, p. 39.

48 Sparrow, 'FDR and the Dust Bowl'.

49 Clay Risen, 'Rightful Heritage: Franklin D. Roosevelt and the Land of America, by Douglas Brinkley', *New York Times*, 23 March 2016, https://www.nytimes.com/2016/03/27/books/review/rightful-heritage-franklin-d-roosevelt-and-the-land-of-america-by-douglas-brinkley.html; Jeremy Deaton, 'How FDR Fought Climate Change: He planted

3 billion trees', published in HuffPost, 7 December 2017, https://nex usmedianews.com/how-fdr-fought-climate-change-d81eee7b1fe1.

50 Henderson, *Letters from the Dust Bowl*.

51 Ibid., p. 167, photo p. 128.

II CELEBRITY FOOD VILLAINS

1 USDA – Farm Service Agency, ARC/PLC Program, http://www.fsa. usda.gov/programs-and-services/arcplc_program/index.

2 Anne Weir Schechinger and Craig Cox, 'Is Federal Crop Insurance Policy Leading to Another Dust Bowl?', EWG, March 2017, https:// cdn.ewg.org/sites/default/files/u352/EWG_DustBowlReport_C07. pdf?_ga=2.51699438.635899846.1606483690-2099343448.1606483690.

3 European Commission, 'Structural Reforms', http://ec.europa. eu/economy_finance/structural_reforms/sectoral/agriculture/ index_en.htm; European Commission, 'The common agricultural policy at a glance: Aims of the common agricultural policy', https://ec.europa.eu/info/food-farming-fisheries/key-policies/ common-agricultural-policy/cap-glance_en.

4 European Commission, *The EU Explained: Agriculture*, Publications Office of the European Union: Luxembourg, November 2014, http:// europa.eu/pol/pdf/flipbook/en/agriculture_en.pdf.

5 European Commission, 'The common agricultural policy (CAP) and agriculture in Europe – Frequently asked questions – Farming in Europe – an overview', 26 June 2013, http://europa.eu/rapid/press-release_MEMO-13-631_en.htm.

6 Calculation based on Cassidy et al., 'Redefining agricultural yields: from tonnes to people nourished per hectare', *Environ. Res. Lett.* 8, 2013, 034015 (8pp), which states that 9:46 x 1015 calories available in plant form are produced by crops globally doi:10.1088/1748-9326/8/3/034015Vaclav Smil, *Feeding the World: A Challenge for the Twenty-First Century*, MIT Press, Cambridge, Massachusetts, 2000; J. Lundqvist, C. de Fraiture and D. Molden, 'Saving Water: From Field to Fork – Curbing Losses and Wastage in the Food Chain', Stockholm International Water Institute Policy Brief, 2008); C. Nellemann, M. MacDevette et al., 'The environmental food crisis – The environment's role in averting future food crises', A UNEP rapid response assessment, United Nations Environment Programme, GRID-Arendal, 2009, www.unep.org/pdf/foodcrisis_lores.pdf.

7 Charoen Pokphand Foods (CPF), https://www.cpfworldwide.com/en/about, accessed 27 April 2021.

8 Cargill 2019 Annual Report, 'Higher Reach', https://www.cargill.com/doc/1432144962450/2019-annual-report.pdf.

9 Tyson Annual Report, *Form 10-K for fiscal year ended September 2019*, United States Securities and Exchange Commission: Washington DC, 2019, https://s22.q4cdn.com/104708849/files/doc_financials/2019/ar/dcdf2f5b-689d-4520-afd6-69691cf580de.pdf.

10 JBS website information, Corporate Profile, accessed June 2021, https://ri.jbs.com.br/en/jbs/corporate-profile/.

11 JBS, 'Annual and Sustainability Report 2019', https://api.mziq.com/mzfilemanager/v2/d/043a77e1-0127-4502-bc5b-21427b991b22/41de5cc6-19dd-a604-4cc3-89450a520625?origin=1.

12 Nutrien 2018 Annual Report, https://www.nutrien.com/sites/default/files/uploads/2019-03/Nutrien_2018_Annual_Report_Enhanced.pdf.

13 Yara Annual Report 2018, https://www.yara.com/siteassets/investors/057-reports-and-presentations/annual-reports/2018/yara-annual-report-2018-web.pdf/.

14 Mosaic Company Annual Report 2018, 'Financial Highlights', accessed February 2020.

15 Nutrien, 2018 Annual Report (accessed February 2020).

16 Corteva Agriscience, 'Factsheet', 2019, https://s23.q4cdn.com/505718284/files/doc_downloads/feature_content/2019/Corteva_FactSheet_9.13.19.pdf.

17 Bayer, Annual report, 2018, https://www.bayer.com/sites/default/files/2020-04/bayer_ar18_entire.pdf.

18 Syngenta Global, 'Company', https://www.syngenta.com/company.

19 Statista, Chemicals and Resources, Chemical Industry, 'BASF's revenue in the Agricultural Solutions segment from 2010 to 2020', https://www.statista.com/statistics/263542/basf-agricultural-solutions-segment-revenue/#:~:text=Revenue%20in%20the%20Agricultural%20Solutions%20segment%20at%20BASF%202009%2D2019&text=In%202019%2C%20BASF%20achieved%20some,in%20the%20Agricultural%20Solutions%20segment.

20 Alliance to Save Our Antibiotics, 'Antibiotic Overuse in Livestock Farming', https://www.saveourantibiotics.org/the-issue/antibiotic-overuse-in-livestock-farming/#:~:text=Worldwide%20it%20is%20estimated%20that,conditions%20where%20disease%20spreads%20easily.

21 Zoetis 2018 Annual Report, https://s1.q4cdn.com/446597350/files/
    doc_financials/2019/ar/Zoetis_2018_Annual_Report.pdf. 'Zoetis
    at a glance', http://www.zoetis.com/about-us/zoetis-at-a-glance.
    aspx; 'Octogain45, Ractopamine Hydrochloride', Zoetis website,
    https://www.zoetisus.com/products/beef/actogain-45.aspx; 'Global
    Manufacturing and Supply', Zoetis website, https://www.zoetis.
    co.uk/global-manufacturing-and-supply.aspx; Merck & Co. Inc.
    2018 Annual Report, *Form 10-K for fiscal year ended December 2018*,
    United States Securities and Exchange Commission: Washington
    DC,   2018,   https://s21.q4cdn.com/488056881/files/doc_financi
    als/2018/Q4/2018-Form-10-K-(without-Exhibits)_FINAL_022
    719.pdf; Elanco 2018 Annual Report, https://s1.q4cdn.com/466533
    431/files/doc_financials/annual/2018-Annual-Report-Final.pdf;
    Boehringer Ingelheim 2018 Annual Report, https://annualreport.
    boehringer-ingelheim.com/fileadmin/downloads/archiv/en/bi_a
    r2018_gesamt_en.pdf; Virbac Group Annual Report 2019, https://
    corporate.virbac.com/files/live/sites/virbac-corporate/files/contribu
    ted/ra2019/Annual_report_2019.pdf; Bayer 2018 Annual report,
    https://www.bayer.com/sites/default/files/2020-04/bayer_ar18_
    entire.pdf.
22 Philip Lymbery, 'Don't all cows eat grass? Part 2: A better way',
    Compassion in World Farming, 2 June 2017, https://www.ciwf.org.
    uk/philip-lymbery/blog/2017/06/dont-all-cows-eat-grass-part-2-a-
    better-way.
23 Agricology, Farmer Profile, 'Neil Heseltine, Hill Top Farm, Malham,
    North Yorkshire', https://www.agricology.co.uk/field/farmer-profi
    les/neil-heseltine.
24 RCVS, 'Dominic Dyer Biography', https://www.rcvs.org.uk/
    who-we-are/vn-council/vn-council-members/Appointed+lay+memb
    ers/dominic-dyer/.
25 Crop Protection Association, 'Who We Are', https://cropprotection.
    org.uk/who-we-are/.
26 Global Consultation Report of the Food and Land Use
    Coalition: Executive Summary, 'Growing Better: Ten Critical
    Transitions to Transform Food and Land Use', Food and Land Use
    Coalition, September 2019, https://www.foodandlandusecoalition.
    org/wp-content/uploads/2019/09/FOLU-GrowingBetter-Globa
    lReport-ExecutiveSummary.pdf.

27 Damian Carrington, '$1m a minute: the farming subsidies destroying
    the world – report', *Guardian*, 16 September 2019, https://www.theg
    uardian.com/environment/2019/sep/16/1m-a-minute-the-farming-
    subsidies-destroying-the-world.

28 International Assessment of Agricultural Knowledge, Science and
    Technology for Development (IAASTD), Executive Summary
    of the Synthesis Report, *Agriculture at a Crossroads* IAASTD,
    Island   Press,   2009,   https://wedocs.unep.org/bitstream/han
    dle/20.500.11822/7880/-Agriculture%20at%20a%20crossroads%20
    -%20Executive%20Summary%20of%20the%20Synthesis%20Rep
    ort-2009Agriculture_at_Crossroads_Synthesis_Report_Execut
    ive_Summary.pdf; Nienke Beintema et al., 'Global Summary for
    Decision Makers', IAASTD, 2009, https://www.researchgate.net/
    publication/269395240_Global_Summary_for_Decision_Makers_
    International_Assessment_of_Agricultural_Science_and_Technol
    ogy_for_Development/link/5489c85c0cf214269f1abc00/download.

29 Millennium Institute, 'Hans Herren, President', https://www.mil
    lennium-institute.org/team/Hans-Herren.

30 Catherine Badgley et al., 'Organic agriculture and the global food
    supply', *Renewable Agriculture and Food Systems*, 2007, 2, 86–108,
    https://doi.org/10.1017/S1742170507001640.

31 FAO, ITPS, GSBI, CBD and EC, *State of Knowledge of Soil
    Biodiversity – Status, challenges and potentialities*, FAO, Rome, 2020),
    p. 192, https://doi.org/10.4060/cb1928en.

12 PANDEMICS ON OUR PLATE

 1 Kevin Liptak and Kaitlan Collins, 'Trump warns of "painful" two
    weeks ahead as White House projects more than 100,000 coronavirus
    deaths', CNN, 31 March 2020, https://edition.cnn.com/2020/03/31/
    politics/trump-white-house-guidelines-coronavirus/index.html.

 2 'Coronavirus: The world in lockdown in maps and charts', BBC News
    World, 7 April 2020, https://www.bbc.co.uk/news/world-52103747.

 3 News Wires, 'UN chief says coronavirus worst global crisis since
    World War II', France24, 1 April 2020, https://www.france24.
    com/en/20200401-un-chief-says-coronavirus-worst-global-cri
    sis-since-world-war-ii.

 4 Martin Bagot and Oliver Milne, 'Boris urged to go on coronavirus
    "war footing" as illness claims first Brit', *Mirror*, 28 February 2020,

https://www.mirror.co.uk/news/uk-news/boris-johnson-urged-go-coronavirus-21601342.

5 Danielle Sheridan, 'Matt Hancock tells Britons we are fighting a war against an "invisible killer" as social distancing measures introduced', *Telegraph*, 16 March 2020, https://www.telegraph.co.uk/polit ics/2020/03/16/coronavirus-cobra-meeting-boris-johnson-chris-whi tty-patrick/.

6 Zoe Drewett, ' "No risk" of catching coronavirus on the Tube, says Sadiq Khan', *Metro*, 3 March 2020, https://metro.co.uk/2020/03/03/ coronavirus-london-tube-sadiq-khan-12339239/.

7 'Coronavirus: Prime Minister Boris Johnson tests positive', BBC News Home, 27 March 2020, https://www.bbc.co.uk/news/ uk-52060791.

8 News Wires, 'UN chief says coronavirus worst global crisis since World War II'.

9 Centers for Disease Control and Prevention, 'Covid Data Tracker: United States COVID-19 Cases, Deaths, and Laboratory Testing (RT-PCR) by State, Territory, and Jurisdiction', https:// www.cdc.gov/coronavirus/2019-ncov/cases-updates/summary. html?CDC_AA_refVal=https%3A%2F%2Fwww.cdc.gov%2Fcoro navirus%2F2019-ncov%2Fsummary.html.

10 Sarah Boseley, 'Calls for global ban on wild animal markets amid coronavirus outbreak', *Guardian*, 24 January 2020, https://www.theg uardian.com/science/2020/jan/24/calls-for-global-ban-wild-animal-markets-amid-coronavirus-outbreak.

11 Ibid.

12 Michael Standaert, 'Coronavirus closures reveal vast scale of China's secretive wildlife farm industry', *Guardian*, 25 February 2020, https:// www.theguardian.com/environment/2020/feb/25/coronavirus-closu res-reveal-vast-scale-of-chinas-secretive-wildlife-farm-industry.

13 Ibid.

14 FAOSTAT, http://www.fao.org/faostat/en/#data/QL.

15 WHO, 'Cumulative number of confirmed human cases for avian influenza A(H5N1)', reported to WHO, 2003–11), https://www.who. int/influenza/human_animal_interface/EN_GIP_20111010Cumula tiveNumberH5N1cases.pdf?ua=1.

16 D. MacKenzie, 'Five easy mutations to make bird flu a lethal pandemic', *New Scientist*, 24 September 2011.

17 'Update: Novel Influenza A (H1N1) Virus Infection – Mexico, March–May, 2009', *MMWR Weekly*, CDC, 5 June 2009, https://www.cdc.gov/mmwr/preview/mmwrhtml/mm5821a2.htm.

18 C. Fraser et al., WHO Rapid Pandemic Assessment Collaboration, 'Pandemic potential of a strain of influenza A (H1N1): early findings', *Science*, 19 June 2009, 324(5934): pp. 1557–61; Y.H. Hsieh et al., 'Early outbreak of 2009 influenza A (H1N1) in Mexico prior to identification of pH1N1 virus', *PLoS One*, 2011, 6(8):e23853, https://www.ncbi.nlm.nih.gov/pmc/articles/PMC3166087/; S. Hashmi, 'La Gloria, Mexico: the possible origins and response of a worldwide H1N1 flu pandemic in 2009', *Am J Disaster Med*, Winter 2013, 8(1): pp. 57–64, https://pubmed.ncbi.nlm.nih.gov/23716374/.

19 Centers for Disease Control and Prevention, 'H1N1 Flu, The 2009 H1N1 Pandemic: Summary Highlights, April 2009–April 2010', updated 16 June 2010, https://www.cdc.gov/h1n1flu/cdcresponse.htm.

20 NHS, 'Swine flu (H1N1)', https://www.nhs.uk/conditions/swine-flu/.

21 Granjas Carroll de Mexico, 'About Us', https://granjascarroll.com/quienes-somos/.

22 Centers for Disease Control and Prevention, 'Influenza (Flu), 2009 H1N1 Pandemic (H1N1pdm09 virus)', https://www.cdc.gov/flu/pandemic-resources/2009-h1n1-pandemic.html.

23 Aaron Kandola, 'Coronavirus cause: Origin and how it spreads', Medical News Today, updated 30 June 2020, https://www.medicalnewstoday.com/articles/coronavirus-causes#origin.

24 Ana Sandoiu, 'Coronavirus: Pangolins may have spread the disease to humans', Medical News Today, 11 February 2020, https://www.medicalnewstoday.com/articles/coronavirus-pangolins-may-have-spread-the-disease-to-humans#How-could-pangolins-have-spread-the-virus?.

25 National Institute of Allergy and Infectious Diseases, 'Coronaviruses', https://www.niaid.nih.gov/diseases-conditions/coronaviruses.

26 Federal Ministry for the Environment, Nature Conservation and Nuclear Safety, 'Minister Schulze: Global nature conservation can reduce risk of future epidemics', BMU, 2 April 2020, https://www.bmu.de/en/pressrelease/minister-schulze-global-nature-conservation-can-reduce-risk-of-future-epidemics/.

27 Larry Light, 'Chinese Virus Could Be a "Black Swan Like No Other": Moody's', Chief Investment Officer, 31 January 2020,

https://www.ai-cio.com/news/chinese-virus-black-swan-like-no-moodys-says/.

28 David Quammen, 'We Made the Coronavirus Epidemic', *New York Times*, 28 January 2020, https://www.nytimes.com/2020/01/28/opin ion/coronavirus-china.html.

29 Damian Carrington, 'Coronavirus: "Nature is sending us a message", says UN Environment Chief', *Guardian*, 25 March 2020, https:// www.theguardian.com/world/2020/mar/25/coronavirus-nature-is-sending-us-a-message-says-un-environment-chief.

30 Damian Carrington, 'Coronavirus UK lockdown causes big drop in air pollution', *Guardian*, 27 March 2020, https://www.theg uardian.com/environment/2020/mar/27/coronavirus-uk-lockd own-big-drop-air-pollution.

31 Jonathan Watts and Niko Kommenda, 'Coronavirus pandemic leading to huge drop in air pollution', *Guardian*, 23 March 2020, https://www.theguardian.com/environment/2020/mar/23/coronavi rus-pandemic-leading-to-huge-drop-in-air-pollution.

32 Charles Riley, '"This is a crisis": Airlines face $113 billion hit from the coronavirus', CNN Business, 6 March 2020, https://edition.cnn. com/2020/03/05/business/airlines-coronavirus-iata-travel/index. html.

33 Watts and Kommenda, 'Coronavirus pandemic leading to huge drop in air pollution'.

34 Gunnar Hökmark, 'Macron shows the "Juncker dilemma" does not exist', EurActiv, 21 June 2017), https://www.euractiv.com/ section/economy-jobs/opinion/macron-shows-the-juncker-dile mma-does-not-exist/.

35 YouTube, 'Keynote: "Peace with Nature: the challenge of sustainability", Karl Falkenberg', Compassion in World Farming, 3 November 2017, https://www.youtube.com/watch?v=3ozm wQ2Pr6g&list=PL-7iZXkicZxfRMp9U7euR3GvhpZTR1 V5y&index=23&t=0s.

13 BUSINESS WITH BOUNDARIES

1 People Pill, 'Petter Stordalen', https://peoplepill.com/people/petter-stordalen/.

2 Walter Willett et al., 'Food in the Anthropocene: the EAT–*Lancet* Commission on healthy diets from sustainable food systems', *Lancet*,

16 January 2019, https://www.thelancet.com/journals/lancet/article/
PIIS0140-6736(18)31788-4/fulltext.

3 Brent Loken et al., 'Diets for a better future', EAT, 2020, https://
eatforum.org/content/uploads/2020/07/Diets-for-a-Better-Future_
G20_National-Dietary-Guidelines.pdf.

4 Stockholm Resilience Centre, 'The nine planetary boundaries',
https://www.stockholmresilience.org/research/planetary-boundar
ies/planetary-boundaries/about-the-research/the-nine-planetary-
boundaries.html.

5 Ibid.

6 'Transformation Towards Planetary Health, Prof. Rockström & Prof.
Jessica Fanzo', EAT Forum, 12 June 2019, https://www.youtube.
com/watch?v=akVONkSdBCQ&list=PLCuQknRNIH2Fvpa0RaL
Qam6Thx-ou3H7H&index=9.

7 International Renewable Energy Agency, 'World Economic Forum
and IRENA Partner for Sustainable Energy Future', 23 September
2020, https://www.irena.org/newsroom/pressreleases/2020/Sep/
WEF-and-IRENA-Partner-for-Sustainable-Energy-Future; Energy
Transitions Commission, 'A global coalition of leaders from across
the energy landscape committed to achieving net-zero emissions
by mid-century', https://www.energy-transitions.org/; Roberto
Bocca and Harsh Vijay Singh, 'The moment of truth for global
energy transition is here', World Economic Forum, 13 May 2020,
https://www.weforum.org/agenda/2020/05/global-energy-transit
ion-index-eti-disrupted-by-covid19/; 'Fostering Effective Energy
Transition', World Economic Forum Insight Report 2020 Edition,
13 May 2020, https://www.weforum.org/reports/fostering-effect
ive-energy-transition-2020#:~:text=The%20Energy%20Transit
ion%20Index%20(ETI,for%20a%20successful%20energy%20transit
ion.&text=Part%20of%20the%20World%20Economic,the%20Ene
rgy%20Architecture%20Performance%20Index.

8 Bocca and Singh, 'The moment of truth for global energy transition
is here'.

9 Hannah Ritchie, 'Food production is responsible for one-quarter
of the world's greenhouse gas emissions', Our World In Data, 6
November 2019, https://ourworldindata.org/food-ghg-emissions.

10 Nicoletta Batini, 'Reaping What We Sow: Smart changes to how we
farm and eat can have a huge impact on our planet', *IMF, Finance and*

Development, December 2019, Vol. 56, No. 4, https://www.imf.org/exter
nal/pubs/ft/fandd/2019/12/farming-food-and-climate-change-batini.
htm.

11 Nicoletta Batini, personal communication, 23 September 2020.

12 Olivier De Schutter, 'Report of the Special Rapporteur on the
    right to food', United Nations General Assembly, 26 December
    2011, Human Rights Council Nineteenth Session, Agenda item
    3, A/HRC/19/59 http://www.ohchr.org/Documents/HRBodies/
    HRCouncil/RegularSession/Session19/A-HRC-19-59_en.pdf.

13 Batini, 'Reaping What We Sow'.

14 'The Greatest Balancing Act: Nature and the Global Economy,
    Based on conservation between David Attenborough and Christine
    Lagarde', *IMF, Finance and Development*, December 2019, Vol. 56,
    No. 4, https://www.imf.org/external/pubs/ft/fandd/2019/12/nature-
    climate-and-the-global-economy-lagarde-attenborough.htm.

15 Doughnut Economics Action Lab, 'About Doughnut Economics',
    https://doughnuteconomics.org/about-doughnut-economics.

16 Daphne Ewing-Chow, 'This new food label will mainstream
    Whole Foods' biggest trend for 2020', *Forbes*, 20 December 2019,
    https://www.forbes.com/sites/daphneewingchow/2019/12/20/
    this-new-food-label-will-mainstream-whole-foods-bigg
    est-trend-for-2020/?sh=53c77fb93933;   Kurt   Knebusch,   'Dig
    the solution: How to offset 100 percent of all greenhouse gas
    emissions', Ohio State University, College of Food, Agricultural and
    Environmental Sciences, 31 July 2015, https://u.osu.edu/sustainabil
    ity/2015/07/31/dig-the-solution-how-to-offset-100-percent-of-all-gre
    enhouse-gas-emissions/.

14 SAVING OURSELVES

1 Ian Redmond, 'What happened to the gorillas that met David
   Attenborough?', BBC Earth, 12 May 2016, http://www.
   bbc.co.uk/earth/story/20160508-what-happened-to-the-
   gorillas-who-met-david-attenborough;   YouTube,   'How   one
   community came together to save the gorilla | Extinction: The
   Facts – BBC', BBC, 5 October 2020, https://www.youtube.com/
   watch?v=_nH_5yjb1bE.

2 Michael Buerk, 'Sir David Attenborough's stark warning in new
   Netflix documentary A Life on Our Planet: "It's about saving

ourselves"', *Radio Times*, 27 September 2020, https://www.radioti mes.com/tv/documentaries/david-attenborough-life-on-our-planet-netflix-big-rt-interview/.

3 David Attenborough, *A Life on Our Planet: My Witness Statement and a Vision for the Future*, Witness Books, London, 2020, p. 7.

4 Worldometer, 'World Population by Year', https://www.worldomet ers.info/world-population/world-population-by-year/.

5 Texas A&M University, 'Humankind did not live with a high-carbon dioxide atmosphere until 1965', ScienceDaily, 25 September 2019, www.sciencedaily.com/releases/2019/09/190925123415.htm.

6 Jon Ungoed-Thomas, '"Healthy" chicken piles on the fat', *The Times*, 3 April 2005, http://www.timesonline.co.uk/printFriendly/0,,1-523-1552131,00.html.

7 United Nations Population Fund, 'World Population Dashboard', https://www.unfpa.org/data/world-population-dashboard; Chelsea Harvey, E&E News, 'CO_2 Levels Just Hit Another Record – Here's Why It Matters', *Scientific American*, 16 May 2019, https://www.scientificamerican.com/article/co2-levels-just-hit-another-rec ord-heres-why-it-matters/.

8 'Global Biodiversity Outlook 5', Secretariat of the Convention on Biological Diversity, Montreal, 2020.

9 Attenborough, *A Life on Our Planet*.

10 R.E.A. Almond, M. Grooten and T. Petersen (eds.), *Living Planet Report 2020: Bending the Curve of Biodiversity Loss*, WWF, Gland, Switzerland, 2020, https://www.wwf.org.uk/sites/defa ult/files/2020-09/LPR20_Full_report.pdf.

11 WWF, 'WWF sends SOS for nature as scientists warn wildlife is in freefall', 9 September 2020, https://www.wwf.org.uk/press-release/ living-planet-report-2020.

12 Yinon M. Bar-On, Rob Phillips and Ron Milo, 'The biomass distribution on Earth', *PNAS*, June 2018, 115(25), pp. 6506–11, https://www.pnas.org/content/115/25/6506.

13 Attenborough, *A Life on Our Planet*, p. 100.

14 Worldometer, 'World Population by Year'.

15 Office for National Statistics, 'UK Environmental Accounts. 2014', 2 July 2014, https://www.ons.gov.uk/economy/environmentalaccou nts/bulletins/ukenvironmentalaccounts/2014-07-02#land-use-exper imental.

16 World Bank, Data, 'Agricultural land (% of land area) – United Kingdom', https://data.worldbank.org/indicator/AG.LND. AGRI.ZS?locations=GB&year_high_desc=false; Department for Environment, Food and Rural Affairs et al., *Agriculture in the United Kingdom 2017*, National Statistics, 2018, https://assets.publishing. service.gov.uk/government/uploads/system/uploads/attachment_d ata/file/741062/AUK-2017-18sep18.pdf.

17 Eurostat, Statistics Explained, 'Land use statistics', https://ec.europa. eu/eurostat/statistics-explained/index.php?title=Land_use_statist ics&oldid=507544

18 World Bank, Data, 'Agricultural land (% of land area) – United Kingdom'; Center for Sustainable Systems, University of Michigan, 'US Cities Factsheet', http://css.umich.edu/factsheets/us-cities-factsh eet#:~:text=Urban%20land%20area%20is%20106%2C386,more%20t han%20double%20by%202060.&text=The%20average%20populat ion%20density%20of,90%20people%20per%20square%20mile.

19 NFU, 'Self-sufficiency Day: Farming growth plan needed', 7 August 2014, https://www.nfuonline.com/self-sufficiency-day-farming-growth-plan-needed/.

20 M. Springmann et al., 'Health and nutritional aspects of sustainable diet strategies and their association with environmental impacts: a global modelling analysis with country-level detail', *Lancet Planet Health*, October 2018, 2(10): e451-e461, doi: 10.1016/S2542-5196(18)30206-7. PMID: 30318102; PMCID: PMC6182055.

21 David R. Williams et al., 'Proactive conservation to prevent habitat losses to agricultural expansion', *Nature Sustainability*, 2020, DOI: 10.1038/s41893-020-00656-5.

22 UNEP, '#FridayFact: Every minute we lose 23 hectares of arable land worldwide to drought and desertification', 12 February 2018, accessed January 2021, https://www.unenvironment.org/news-and-stories/ story/fridayfact-every-minute-we-lose-23-hectares-arable-land-worldwide-drought; FAOSTAT, 'Land use', http://www.fao.org/faos tat/en/#data/RL.

23 *National Geographic*, 'How to live with it: Crop changes', https://www. nationalgeographic.com/climate-change/how-to-live-with-it/crops. html; Alessandro De Pinto et al., 'Climate-smart agriculture and global food-crop production', *PLoS ONE*, 2020, 15(4): e0231764, https://doi. org/10.1371/journal. pone.0231764.

24 FAO, ITPS, GSBI, SCBD and EC, *State of Knowledge of Soil Biodiversity – Status, challenges and potentialities*, FAO: Rome, 2020, https://doi.org/10.4060/cb1929en, http://www.fao.org/3/cb1929en/ CB1929EN.pdf.

25 Attenborough, *A Life on Our Planet*, p. 161.

26 University of Bristol, 'Recovering From A Mass Extinction', ScienceDaily, 20 January 2008, http://www.sciencedaily.com/relea ses/2008/01/080118101922.htm.

27 Attenborough, *A Life on Our Planet*, p. 6.

28 C.B. Field et al. (eds.), 'Summary for policymakers', in *Climate Change 2014: Impacts, Adaptation, and Vulnerability, Part A: Global and Sectoral Aspects*, Contribution of Working Group II to the Fifth Assessment Report of the Intergovernmental Panel on Climate Change, Cambridge University Press, Cambridge, UK, and New York, USA, 2014), pp. 1–32, http://www. ipcc.ch/pdf/assessment-report/ar5/wg2/ar5_wgII_spm_en.pdf.

29 Andy Challinor et al., *Climate and global crop production shocks*, Resilience Taskforce Sub Report, Annex A (Global Food Security Programme: UK, 2015), https://www.foodsecurity.ac.uk/ publications/resilience-taskforce-sub-report-annex-climate-global-crop-production-shocks.pdf.

30 M. Le Page, 'US cities to sink under rising seas', *New Scientist*, 17 October 2015, Vol. 228, Issue 3043, p. 8; M. Le Page, 'Even drastic emissions cuts can't save New Orleans and Miami', *New Scientist*, 14 October 2015, https://www.newscientist.com/article/mg22830433-900-even-drastic-emissions-cuts-cant-save-new-orleans-and-miami/; B.H. Strauss, S. Kulp & A. Levermann, 'Carbon choices determine US cities committed to futures below sea level', *PNAS*, 2015, early edition', Vol. 112, no. 44, www. pnas.org/cgi/doi/10.1073/pnas.1511186112.

31 Attenborough, *A Life on Our Planet*, p. 105.

32 Ibid., p. 164.

33 Ibid., p. 170.

34 Ibid., p. 171.

35 Buerk, 'Sir David Attenborough's stark warning'.

15 REGENERATION

1 Douglas C. Munski, Bernard O'Kelly and Elwyn B. Robinson, 'North Dakota', *Encyclopedia Britannica*, 29 October 2020, accessed 13 May 2021, https://www.britannica.com/place/North-Dakota.

2 Brown's Ranch, 'Cropping', http://brownsranch.us/cropping/.

3 Gabe Brown, *Dirt to Soil: One Family's Journey into Regenerative Agriculture*, Chelsea Green, Vermont, USA, 2018, p. 109.

4 Ibid., p. 21.

5 Ibid., p. 22.

6 Ibid.

7 Ibid., p. 13.

8 Brown's Ranch, 'Cropping'.

9 Brown, *Dirt to Soil*, p. 68.

10 Ibid., p. 89.

11 Brown's Ranch, 'Food', http://brownsranch.us/food/.

12 Brown, *Dirt to Soil*, p. 119.

13 Ibid., p. 3.

14 W.R. Teague et al., 'Grazing management impacts on vegetation, soil biota and soil chemical, physical and hydrological properties in tall grass prairie', *Agriculture, Ecosystems & Environment*, 2011, 141, pp. 310–22; W.R. Teague et al., 'The role of ruminants in reducing agriculture's carbon footprint in North America', *Journal of Soil and Water Conservation*, 2016, 71, pp. 156–64.

15 David R. Montgomery, *Growing a Revolution: Bringing our Soil Back to Life*, WW Norton & Company, New York, 2017, p. 194.

16 Brown, *Dirt to Soil*, p. 119.

17 Ibid., p. 86.

18 Ibid.

19 Brown's Ranch, 'Food'.

20 Ibid.

21 Gabe Brown, personal communication, 20 April 2021.

22 Brown, *Dirt to Soil*, pp. 122, 130.

23 Brown's Ranch, 'Home: Welcome to Brown's Ranch', http://brownsranch.us/.

24 YouTube, 'PFLA Webinar: Pastured Pigs – Maximising Forage Feeding', Pasture Fed Livestock Association, 29 May 2020, https://www.youtube.com/watch?v=y-tBDsjYTsY.

25 YouTube, 'How Dairy Farming Is Becoming More Ethical', Farmdrop, 23 October 2019, https://www.youtube.com/watch?v=yNRpjuixmYQ.

26 Tom Philpott, 'One weird trick to fix farms forever', *Mother Jones*, 9 September 2013, https://www.motherjones.com/environment/2013/09/cover-crops-no-till-david-brandt-farms/.

27 Brown, *Dirt to Soil*, pp. 52–3.

28 Chris Kick, 'Mimicking nature: Cover crop guru Dave Brandt was an early adapter', *Farm and Dairy*, 8 June 2016, https://www.farmanddairy.com/news/mimicking-nature-cover-crop-guru-dave-brandt-was-an-early-adapter/340579.html.

29 Philpott, 'One weird trick to fix farms forever'.

30 Jennifer Kiel, 'Meet Master Farmer Brandt: Rooted in soil health', *Ohio Farmer*, 1 April 2016, https://www.farmprogress.com/story-meet-master-farmer-brandt-rooted-soil-health-9-139545; Montgomery, *Growing a Revolution*, pp. 235–6.

31 Soil Health Academy, 'Ohio Soil Health Pioneer's Farm is Classroom for Upcoming Regenerative Agriculture School', Regeneration International, 16 May 2019, https://regenerationinternational.org/2019/05/16/ohio-soil-health-pioneers-farm-is-classroom-for-upcoming-regenerative-agriculture-school/#:~:text=Today%2C%20the%20Soil%20Health%20Academy,and%20improved%20his%20farm's%20profitability.

32 Randall Reeder and Rafiq Islam, 'No-till and conservation agriculture in the United States: An example from the David Brandt farm, Carroll, Ohio', *International Soil and Water Conservation Research*, 2014, 2, pp. 97–107, https://doi.org/10.1016/S2095-6339(15)30017-4.

33 Ibid.

34 Kiel, 'Meet Master Farmer Brandt'.

35 Montgomery, *Growing a Revolution*, pp. 230–31.

36 R. Lal, 'Enhancing crop yields in the developing countries through restoration of the soil organic carbon pool in agricultural lands', *Land Degradation and Development*, 2005, https://onlinelibrary.wiley.com/doi/epdf/10.1002/ldr.696.

37 Montgomery, *Growing a Revolution*, p. 234.

38 Ken Roseboro, 'Soil Health: The next agricultural revolution', EcoWatch, 7 January 2019, https://www.ecowatch.com/soil-health-as-the-next-agricultural-revolution-2625362894.html.

39 W.A. Albrecht, *Loss of Soil Organic Matter and Its Restoration*, US Dept of Agriculture, Soils and Men, Yearbook of Agriculture 1938, pp. 347–60, https://soilandhealth.org/wp-content/uploads/01aglibrary/010120albrecht.usdayrbk/lsom.html.

40 Montgomery, *Growing a Revolution*, p. 245.

41 YouTube, 'Growing Underground: The world's first subterranean farm', Growing Underground, 7 August 2015, https://www.youtube.com/watch?v=Co1qpywMHNQ.

42 Growing Underground, http://growing-underground.com/; YouTube, 'How do hydroponics work? Underground Farming', BBC Earth Lab, 8 December 2016, https://www.youtube.com/watch?v=FecuxUotMmE.

43 YouTube, 'Growing Underground'.

44 Zlata Rodionova, 'Inside London's first underground farm', *Independent*, 3 February 2017, https://www.independent.co.uk/Business/indyventure/growing-underground-london-farm-food-waste-first-food-miles-a7562151.html.

45 Sophia Epstein, 'Growing underground: the hydroponic farm hidden 33 metres below London', *Wired*, 13 April 2017, https://www.wired.co.uk/article/underground-hydroponic-farm.

46 Rodionova, 'Inside London's first underground farm'.

47 Epstein, 'Growing underground'.

48 AeroFarms, 'AeroFarms® to build world's largest R&D Indoor Vertical Farm in Abu Dhabi as part of USD $100 million AgTech investment by Abu Dhabi Investment Office (ADIO)', 9 April 2020, https://aerofarms.com/2020/04/09/aerofarms-to-build-worlds-largest-rd-farm/#:~:text=Newark%2C%20NJ%20USA%20%E2%80%93%20April%209,world's%20largest%20of%20its%20kind.

49 Malavika Vyawahare, 'World's largest vertical farm grows without soil, sunlight or water in Newark', *Guardian*, 14 August 2016, https://www.theguardian.com/environment/2016/aug/14/world-largest-vertical-farm-newark-green-revolution.

50 Ibid.

16 RETHINKING PROTEIN

1 Y.N. Harari, 'Extinction and Livestock Conference – Video contribution', Compassion in World Farming and WWF-UK, London, 6 October 2017.

2 Vegan Society, News/Market Insights, 'Plant Milk Market', https://www.vegansociety.com/news/market-insights/plant-milk-market#:~:text=In%20terms%20of%20global%20revenue,16.7%25%20between%202020%20%E2%80%93%202025; Innova Market

Insights, 'Global Plant Milk Market to Top US $16 Billion in 2018: Dairy Alternative Drinks Are Booming, Says Innova Market Insights', PR Newswire, 13 June 2017, https://www.prnewswire. com/news-releases/global-plant-milk-market-to-top-us-16-bill ion-in-2018--dairy-alternative-drinks-are-booming-says-innova-mar ket-insights-300472693.html.

3 Good Food Institute, 'US retail market data for the plant-based industry', accessed June 2021, https://gfi.org/marketresearch/.

4 Mintel, 'Milking the vegan trend: A quarter (23%) of Brits use plant-based milk', Mintel Food and Drink, 19 July 2019, https://www.min tel.com/press-centre/food-and-drink/milking-the-vegan-trend-a-quarter-23-of-brits-use-plant-based-milk.

5 Packaged Facts, '5 Dairy Alternative Beverage Trends to Watch in 2018', PR Newswire, 2 November 2017, https://www.prn ewswire.com/news-releases/5-dairy-alternative-beverage-tre nds-to-watch-in-2018-300548626.html.

6 Ibid.

7 L. Rojas, 'International Pesticide Market and Regulatory Profile', Worldwide Crop Chemicals, http://wcropchemicals.com/pesticide _regulatory_profile/#_ftn28.

8 L.M. Bombardi, interviewed by the author at the University of São Paulo, Brazil, 4 March 2016.

9 Ian Johnston, 'Industrial farming is driving the sixth mass extinction of life on Earth, says leading academic', Independent, 27 August 2017, http:// www.independent.co.uk/environment/mass-extinction-life-on-earth-farming-industrial-agriculture-professor-raj-patel-a7914616.html.

10 W. Fraanje and T. Garnett, 'Soy: food, feed, and land use change', Foodsource: Building Blocks, Food Climate Research Network, University of Oxford, 30 January 2020, https://www.leap.ox.ac.uk/ article/soy-food-feed-and-land-use-change.

11 WWF Global, 'The growth of soy, impacts and solutions: The market for soy in europe', WWF Global, 2014, http://wwf.panda.org/what_ we_do/footprint/agriculture/soy/soyreport/the_continuing_rise_of_ soy/the_market_for_soy_in_europe/.

12 Alpro, accessed 4 June 2021, https://www.alpro.com/uk/good-for-the-planet/; Alpro, 'Fact Sheet: First Flemish soya soon to be harvested', https://www.alpro.com/upload/press/misc/fact-sheet-first-flem ish-soya-soon-to-be-harvested.pdf.

13 S. Birgersson, B.S. Karlsson and L. Söderlund, 'Soy milk, an attributional life cycle assessment examining the potential environmental impact of soy milk', Group LCA04 Project Report – Life Cycle Assessment AG2800, Stockholm, May 2009, http://envo rmation.org/wp-content/uploads/2015/08/Soy-milk-an-attributio nal-Life-Cycle-Assessment-examining-the-potential-environmen tal-impact-of-soy-milk.pdf,____https://www.academia.edu/31017783/ Soy_Milk_an_attributional_Life_Cycle_Assessment_examining_ the_potential_environmental_impact_of_soy_milk.

14 Institute of the Environment and Sustainability, 'Moving science to action', https://www.ioes.ucla.edu/wp-content/uploads/cow-vs-alm ond-milk-1.pdf.

15 Almond Breeze, 'Where do your almonds come from?', https://www. almondbreeze.co.uk/faq/.

16 P. Sullivan, Business Development EU, Blue Diamond Almonds, in response to customer service enquiry, 24 January 2018.

17 Alpro, 4 June 2021.

18 Tom Levitt, '"Wow, no cow": the Swedish farmer using oats to make milk', *Guardian*, 26 August 2017, https://www.theguardian. com/sustainable-business/2017/aug/26/wow-no-cow-swedish-far mer-oats-milk-oatly.

19 Oatly, 'The climate footprint of enriched oat drink ambient', CarbonCloud AB, Sweden, 2019, https://www.oatly.com/uploads/ attachments/ck16jh9jt04k9bggixfg6ssrn-report-the-climate-footpr int-of-enriched-oat-drink-ambient-carboncloud-20190917.pdf; S. Clune, E. Crossin and K. Verghese, 'Systematic review of greenhouse gas emissions for different fresh food categories', *Journal of Cleaner Production*, 2017, 140, pp. 766–83; SIK AB, Swedish Institute for Food and Biotechnology, on behalf of Oatly AB, 'Life cycle assessment summary', copy provided by company.

20 Tom Levitt, 'Animal-free dairy products move a step closer to market', *Guardian*, 13 September 2016, https://www.theguardian.com/envi ronment/2016/sep/13/animal-free-dairy-products-move-a-step-clo ser-to-market.

21 Zsofia Mendly-Zambo, Lisa Jordan Powell and Lenore L. Newman, 'Dairy 3.0: Cellular agriculture and the future of milk', *Food, Culture & Society*, 2021, 24:5, pp. 675–93, https://www.tandfonline.com/doi/ full/10.1080/15528014.2021.1888411.

22 Cathy Owen, 'The goats that are taking over a Welsh seaside town during lockdown', Wales Online, 31 March 2020, https://www.walesonline.co.uk/news/wales-news/goats-taking-over-welsh-seaside-18011581.

23 Hotels.com, '10 best things to do in Llandudno', https://uk.hotels.com/go/wales/things-to-do-llandudno.

24 Nathaniel Bullard, 'The World Is Finally Losing Its Taste for Meat', Bloomberg, 30 July 2020, https://www.bloomberg.com/news/artic les/2020-07-30/good-news-for-climate-change-as-world-loses-it s-taste-for-meat?utm_campaign=likeshopme&utm_medium=instag ram&sref=aGTrSb9U&utm_source=dash%20hudson&utm_cont ent=www.instagram.com/p/CDS0AKcn8Yf/.

25 Justin McCarthy, 'Nearly one in four in US have cut back on eating meat', Gallup, 27 January 2020, https://news.gallup.com/poll/282 779/nearly-one-four-cut-back-eating-meat.aspx.

26 Rebecca Smithers, 'Third of Britons have stopped or reduced eating meat – Report', Guardian, 1 November 2018, https://www.theguard ian.com/business/2018/nov/01/third-of-britons-have-stopped-or-reduced-meat-eating-vegan-vegetarian-report.

27 Melissa Clark, 'The Meat-Lover's Guide to Eating Less Meat', New York Times, 31 December 2019, https://www.nytimes.com/2019/12/31/din ing/flexitarian-eating-less-meat.html.

28 Fox News, 'Arnold Schwarzenegger joins celebrities claiming eating less meat will help climate change', 10 December 2015, last updated 20 March 2018, https://www.foxnews.com/food-drink/arnold-schwarzenegger-joins-celebrities-claiming-eating-less-meat-will-help-climate-change.

29 Roger Harrabin, 'COP21: Arnold Schwarzenegger: "Go part-time vegetarian to protect the planet"', BBC News, 8 December 2015, https://www.bbc.co.uk/news/science-environment-35039465.

30 SIVO Insights, 'Marketing to Food Tribes', https://sivoinsights. com/2015/10/marketing-to-food-tribes/.

31 Emiko Terezono and Leslie Hook, 'Have we reached "peak meat"?', Financial Times, 26 December 2019, https://www.ft.com/cont ent/815c9d62-14f4-11ea-9ee4-11f260415385.

32 FAO, 2020 food outlook: biannual report on global food markets, FAO, Rome, June 2020, Food Outlook 1, https://doi.org/10.4060/ ca9509en.

33 Ibid.

34 D. Mason-D'Croz, J.R. Bogard, M. Herrero et al., 'Modelling the global economic consequences of a major African swine fever outbreak in China', *Nature Food*, 2020, 1, pp. 221–8, https://www.nat ure.com/articles/s43016-020-0057-2?draft=marketing.

35 Elaine Watson, 'How is coronavirus impacting plant-based meat? Impossible Foods, Lightlife, Tofurky, Meatless Farm Co, Dr. Praeger's, weigh in', *FoodNavigator*, 6 April 2020, https://www. foodnavigator-usa.com/Article/2020/04/06/How-is-coronavirus-impacting-plant-based-meat-Impossible-Foods-weighs-in.

36 Markets and Markets, 'Plant-based Meat Market: Plant-based meat market by source, product, type, process and region: Global forecast to 2025', December 2020, https://www.marketsandmarkets.com/Market-Reports/plant-based-meat-market-44922705.html.

37 Research and Markets, 'Global meat sector market analysis & forecast report, 2019: A $1.14 trillion industry opportunity by 2023', Globe Newswire, 2 May 2019, https://www.globenewswire.com/news-release/2019/05/02/1815144/0/en/Global-Meat-Sector-Market-Analysis-Forecast-Report-2019-A-1-14-Trillion-Industry-Opportunity-by-2023.html.

38 FAOSTAT database, 'Food supply, new food balances', FAO, accessed October 2021, http://www.fao.org/faostat/en/#data/FBS.

39 Ibid.

40 Public Health England, 'Statistical Summary: National diet and nutrition survey: Years 1 to 9 of the rolling programme (2008/09 – 2016/17): Time trend and income analyses', January 2019, https://ass ets.publishing.service.gov.uk/government/uploads/system/uploads/attachment_data/file/772430/NDNS_Y1-9_statistical_summary. pdf, see also tables 3.8 to 3.16, https://assets.publishing.service.gov. uk/government/uploads/system/uploads/attachment_data/file/772 667/NDNS_UK_Y1-9_Data_tables.zip.

41 Estimated from the yield of edible meat from the average carcass weight of beef cattle and meat chickens.

42 FAOSTAT database, 'Food Supply.

43 Mildred Haley, 'Livestock, dairy and poultry outlook', USDA, 19 January 2018, https://www.ers.usda.gov/publications/pub-deta ils/?pubid=86848.

44 Linn Åkesson, 'Historical reduction in meat consumption in Sweden – millions of animals affected', Djurens Rätt, 16 March 2021, https://www.djurensratt.se/blogg/historical-reduction-meat-cons umption-sweden-millions-animals-affected.

45 EC, *EU agricultural outlook for markets and income, 2019-2030*, European Commission, DG Agriculture and Rural Development: Brussels, 2019, Graph 5.5 EU meat consumption kg per cap, https://ec.europa.eu/info/sites/info/files/food-farm ing-fisheries/farming/documents/agricultural-outlook-2019-report _en.pdf.

46 Ibid., p. 44; OECD, 'Meat consumption (indicator)', 2019, accessed on 18 February 2019, https://data.oecd.org/agroutput/meat-cons umption.htm.

47 M. Springmann et al., 'Health and nutritional aspects of sustainable diet strategies and their association with environmental impacts: a global modelling analysis with country-level detail', *Lancet Planet Health*, 2018, Appendix, supplementary information, 2, e451–61.

48 M. Springmann et al., 'Options for keeping the food system within environmental limits', *Nature*, 2018, https://www.nature.com/artic les/s41586-018-0594-0.

49 B. Bajželj et al., 'Importance of food-demand management for climate mitigation', *Nature Climate Change*, October 2014, Vol. 4, http://www.nature.com/doifinder/10.1038/nclimate2353;. M. Springmann et al., 'Analysis and valuation of the health and climate change cobenefits of dietary change', *PNAS*, 2016, 113, 15, pp. 4146–51; Springmann et al., 'Options for keeping the food system within environmental limits'.

50 A. Leip et al., *Evaluation of the livestock sector's contribution to the EU greenhouse gas emissions*, European Commission's Joint Research Centre, 2019.

51 E. Wollenberg et al., 'Reducing emissions from agriculture to meet the 2°C target', *Global Change Biology*, 2016, 22, pp. 3859–64.

52 EC, *EU agricultural outlook for markets and income, 2019-2030*.

53 Antony Froggatt and Laura Wellesley, 'Meat analogues: Considerations for the EU', Chatham House: London, 2019, https://reader.chath amhouse.org/meat-analogues-considerations-eu#.

54 Ibid.

55 Chelsea Gohd, 'Meat grown in space for the first time ever', Space. com, 2020, https://www.space.com/meat-grown-in-space-station-bioprinter-first.html.

56 NASA, Webcast, 'STS-111 International Space Station: Question and answer board', NASA's John F. Kennedy Space Center, page last updated 22 November 2007, https://www.nasa.gov/missions/highlig hts/webcasts/shuttle/sts111/iss-qa.html.

57 Rebecca Smithers, 'First meat grown in space lab 248 miles from Earth', *Guardian*, 7 October 2019, https://www.theguardian.com/ environment/2019/oct/07/wheres-the-beef-248-miles-up-as-firs t-meat-is-grown-in-a-space-lab.

58 P.R. Shukla et al. (eds.), *Climate Change and Land: An IPCC Special Report on climate change, desertification, land degradation, sustainable land management, food security and greenhouse gas fluxes in terrestrial ecosystems*, IPCC, 2019, https://www.ipcc.ch/srccl/ https://www.ipcc. ch/srccl/chapter/summary-for-policymakers/.

59 Hanna Tuomisto and M.J. Mattos, 'Life cycle assessment of cultured meat production', 7th International Conference on Life Cycle Assessment in the Agri-Food Sector, Bari, Italy, 22–24 September 2010, https://www.researchgate.net/publication/215666764_Life_ cycle_assessment_of_cultured_meat_production#:~:text=Cultu red%20meat%20is%20produced%20in%20vitro%20by%20us ing%20tissue%20engineering%20techniques.&text=Life%20cy cle%20assessment%20(LCA)%20research,source%20for%20mus cle%20cell%20growth.

60 Elliot Swartz, 'New studies show cultivated meat can have massive environmental benefits and be cost-competitive by 2030', Good Food Institute, 9 March 2021, https://gfi.org/blog/cultivated-meat-lca-tea/.

61 Gohd, 'Meat grown in space for the first time ever'.

62 'KFC wants to make 3D bioprinted chicken nuggets in "restaurant of the future"', *Newsround*, BBC, 21 July 2020, https://www.bbc.co.uk/ newsround/53471685; Kat Smith, 'KFC Is Developing Lab-Grown Chicken Nuggets in Russia', LiveKindly, 19 July 2020, https://www. livekindly.co/kfc-lab-grown-chicken-nuggets-russia/.

63 Sarah Young, '"Meat of the future": KFC is developing the world's first lab grown chicken nuggets', *Independent*, 21 July 2020, https:// www.independent.co.uk/life-style/food-and-drink/kfc-lab-grown-chicken-nuggets-biomeat-3d-bioprinting-russia-a9629671.html.

64 Tor Marie, 'Clean Meat Startups: 10 lab-grown meat producers to watch', TechRound, 23 July 2019, https://techround.co.uk/startups/clean-meat-startups/.

65 Derrick A. Paulo and Chua Dan Chyi, 'Grow meat at home from stem cells? It's coming, says Shiok Meats CEO', CNA Insider, 7 March 2020, https://www.channelnewsasia.com/news/cnainsider/lab-grow-stem-cell-based-protein-home-shiok-meats-sandhya-sriram-12511730.

66 Mosa Meat, 'Growing Beef', https://www.mosameat.com/technology.

67 YouTube, 'Taste test of world's first lab-grown burger that cost £215,000 to produce', Leak Source News, 6 August 2013, https://www.youtube.com/watch?v=9XqcIkbxxBw; Alok Jha, 'First lab-grown hamburgers gets full marks for "mouth feel"', *Guardian*, 6 August 2013, https://www.theguardian.com/science/2013/aug/05/world-first-synthetic-hamburger-mouth-feel.

68 Mosa Meat, 'FAQs', https://www.mosameat.com/faq.

69 Chase Purdy, *Billion Dollar Burger: Inside Big Tech's Race for the Future of Food*, Piatkus, London, 2020, p. xvi.

70 Mosa Meat, 'FAQs'.

71 Cell Based Tech, 'Lab grown meat companies', https://cellbasedtech.com/lab-grown-meat-companies.

72 Isha Datar, personal communication, 28 August 2020.

73 YouTube, 'Is cell-cultured meat ready for the mainstream?', Quartz News, 1 November 2019, https://www.youtube.com/watch?v=VYXw_-vJFBA.

74 Ibid.

75 Good Food Institute, 'Deep Dive: Cultivated meat cell culture media', accessed June 2021, https://gfi.org/science/the-science-of-cultivated-meat/deep-dive-cultivated-meat-cell-culture-media/.

76 YouTube, 'Is cell-cultured meat ready for the mainstream?'.

77 Tor Marie, 'Clean Meat Startups'.

78 Good Food Institute, 'Deep Dive'.

79 Mosa Meat, 'FAQs'.

80 Harry de Quetteville, 'The future of meat: Plant-based, lab-grown and goodbye to the abattoir', *Telegraph*, 16 July 2020, https://www.telegraph.co.uk/food-and-drink/features/future-meat-plant-based-lab-grown-goodbye-abattoir/.

81 Mary Ellen Shoup, 'Survey: How do consumers feel about cell cultured meat and dairy minus the cows?', *FoodNavigator*, 13 December 2019, https://www.foodnavigator-usa.com/Article/2019/12/13/Survey-Are-consumers-warming-up-to-the-idea-of-cell-cultured-meat#.

82 Purdy, *Billion Dollar Burger*, p. 195.

83 Damian Carrington, 'World's first lab-grown steak revealed – but the taste needs work', *Guardian*, 14 December 2018, https://www.theguard ian.com/environment/2018/dec/14/worlds-first-lab-grown-beef-steak-revealed-but-the-taste-needs-work.

84 Catherine Tubb and Tony Seba, *Rethinking Food and Agriculture 2020–2030: The Second Domestication of Plants and Animals, the Disruption of the Cow, and the Collapse of Industrial Livestock Farming*, RethinkX, San Francisco, 2019, https://static1.squarespace.com/sta tic/585c3439be65942f022bbf9b/t/5d7feoe83d119516bfcoo17e/156866 1791363/RethinkX+Food+and+Agriculture+Report.pdf.

85 Brent Huffman, 'Cow', *Encyclopedia Britannica*, 26 November 2019, https://www.britannica.com/animal/cow#ref1242584.

86 Quorn, 'Mycoprotein: Super protein. Super tasty', https://www.quorn.co.uk/mycoprotein.

87 Meati, https://meati.com/pages/what-is-mycelium.

88 Liz Specht and Nate Crosser (principal authors), 'State of the industry report – Fermentation: An introduction to a pillar of the alternative protein industry', Good Food Institute, 2020, https://gfi.org/wp-cont ent/uploads/2021/01/INN-Fermentation-SOTIR-2020-0910.pdf.

89 Tubb and Seba, *Rethinking Food and Agriculture 2020–2030*, pp. 13–14.

90 Ibid., p. 16.

91 Donavyn Coffey, 'New report calls fermentation the next pillar of alternative proteins', The Spoon, 17 September 2020, https://thesp oon.tech/new-report-calls-fermentation-the-next-pillar-of-alternat ive-proteins/.

92 Jeanne Yacoubou, 'Microbial Rennets and Fermentation Produced Chymosin (FPC): How vegetarian are they?', Vegetarian Resource Group Blog, 21 August 2012, https://www.vrg.org/blog/2012/08/21/ microbial-rennets-and-fermentation-produced-chymosin-fpc-how-vegetarian-are-they/; Flora Southey, '"Game changer" cheese enzyme increases yield by up to 1%: "There is nothing on par with this"', *FoodNavigator*, 3 April 2019, https://www.foodnavigator. com/Article/2019/04/03/Game-changer-cheese-enzyme-increa

ses-yield-by-up-to-1-There-is-nothing-on-par-with-this?utm
_source=newsletter_weekly&utm_medium=email&utm_campa
ign=From%2029-Mar-2019%20to%2005-Apr-2019&c=HzAKb
DRqOiSBI6S5LgpeUBGoALkIa557&p2=.

93 *FoodNavigator*, 'Sectors: Meat', http://www.globalmeatnews.com/
Industry-Markets/Study-suggests-conventional-meat-will-cost-a-
premium-as-demand-grows.

94 Edward Devlin, 'At-home demand helps Quorn offset foodservice
shutdown', *The Grocer*, 31 July 2020, https://www.thegrocer.co.uk/
results/at-home-demand-helps-quorn-offset-foodservice-shutd
own/646997.article.

95 Carbon Trust, 'Quorn – product carbon footprinting and labelling',
https://www.carbontrust.com/our-projects/quorn-product-carbon-
footprinting-and-labelling.

96 Sarah Buhr, 'Former jock hatching new food biz with help from
tech', Special for *USA Today*, 2014, https://eu.usatoday.com/story/
tech/2014/04/07/hampton-creek-foods-josh-tetrick-just-mayo-sili
con-valley/7348077/.

97 Purdy, *Billion Dollar Burger*, pp. 49–60; Ruben Baartmay, 'How
the Dutch government is obstructing the advent of in vitro meat',
NextNature.Net, 22 May 2018, https://nextnature.net/story/2018/
interview-ira-van-eelen.

98 Eat Just, Inc., 'Eat Just makes history (again) with restaurant debut
of cultured meat', Business Wire, 21 December 2020, https://www.
businesswire.com/news/home/20201220005063/en/Eat-Just-Makes-
History-Again-with-Restaurant-Debut-of-Cultured-Meat.

99 Ancient History Lists, 'Top 11 Inventions and Discoveries of
Mesopotamia', last updated by Saugat Adhikari, 20 November 2019,
https://www.ancienthistorylists.com/mesopotamia-history/top-11-
inventions-and-discoveries-of-mesopotamia/#1_Agriculture_and
_Irrigation; Time Maps, 'Ancient Mesopotamia: Civilization and
Society', https://www.timemaps.com/civilizations/ancient-meso
potamia/; Megan Gambino, 'A Salute to the Wheel', *Smithsonian
Magazine*, 17 June 2009, https://www.smithsonianmag.com/scie
nce-nature/a-salute-to-the-wheel-31805121/.

100 E.S. Cassidy et al., 'Redefining agricultural yields: from tonnes to
people nourished per hectare', *Environ. Res. Lett.*, 2013, 8 034015,
stacks.iop.org/ERL/8/034015.

101 M. Henchion et al., 'Future Protein Supply and Demand: Strategies
     and Factors Influencing a Sustainable Equilibrium', *Foods*, 20 July
     2017, 6(7), p. 53, https://www.ncbi.nlm.nih.gov/pmc/articles/PMC
     5532560/; FAOSTAT database, 'New Food Balances', http://www.
     fao.org/faostat/en/#data/FBS.

102 J. Poore and T. Nemecek, 'Reducing food's environmental
     impacts through producers and consumers', *Science*, 2018, Vol.
     360, Issue 6392, pp. 987–92, https://science.sciencemag.org/cont
     ent/360/6392/987.

103 Jonathan B. Wight, 'The Stone Age didn't end because we ran
     out of stones', *Economics and Ethics*, 27 March 2014, https://
     www.economicsandethics.org/2014/03/the-stone-age-didnt-end-
     because-we-ran-out-of-stones-.html#:~:text=Friedman%20no
     tes%3A%20%E2%80%9CThe%20former%20Saudi,%2C%20wh
     ich%20were%20more%20productive.%E2%80%9D; Matt Frei,
     'Washington diary: Oil addiction', BBC News, 3 July 2008, http://
     news.bbc.co.uk/1/hi/world/americas/7486705.stm.

104 Information from Drovers website.

105 Shruti Singh, 'Bill Gates and Richard Branson back start-up that
     grows "clean meat"', Bloomberg, 23 August 2017, https://www.
     bloomberg.com/news/articles/2017-08-23/cargill-bill-gates-bet-on-
     startup-making-meat-without-slaughter.

106 Tubb and Seba, *Rethinking Food and Agriculture 2020-2030*, p. 20.

107 Ibid.

108 Ibid.

109 John Greathouse, 'Here's how Lisa Dyson's start-up is reducing
     world hunger and combating climate change', Forbes, 10 March
     2020, https://www.forbes.com/sites/johngreathouse/2020/03/10/
     heres-how-lisa-dysons-startup-is-reducing-world-hunger-and-
     combating-climate-change/#2fd3162f52f9; Air Protein, 'The future
     of meat', https://www.airprotein.com/.

17 REWILDING

1 Keiron Pim, 'I'm the luckiest man I know', *Norwich Evening News*,
   25 July 2011, https://www.eveningnews24.co.uk/views/i-m-the-
   luckiest-man-i-know-1-973427.

2 H. Charles J. Godfray et al., 'Food security: The challenge of feeding 9 billion people', *Science*, 12 February 2010, 327, pp. 812–18, https://science.sciencemag.org/content/327/5967/812.

3 D. Ray, N. Ramankutty, N. Mueller et al., 'Recent patterns of crop yield growth and stagnation', *Nature Communications*, 2012, 3, p. 1293, https://www.nature.com/articles/ncomms2296.

4 Isabella Tree, *Wilding: The return of nature to a British farm*, Picador, London, 2018.

5 Knepp Wild Range Meat, 'Wild Range Meat', https://www.kneppwildrangemeat.co.uk/.

6 Sam Knight, 'Can farming make space for nature?', *New Yorker*, 10 February 2020, https://www.newyorker.com/magazine/2020/02/17/can-farming-make-space-for-nature.

7 'Norfolk four-course system', *Encyclopedia Britannica*, 20 July 1998, https://www.britannica.com/topic/Norfolk-four-course-system.

8 Holkham, 'Crop Rotation', https://www.holkham.co.uk/farming-shooting-conservation/crop-rotation/coke-of-norfolk; Holkham, 'Crop Rotation: Diversity of Cropping', https://www.holkham.co.uk/farming-shooting-conservation/crop-rotation/six-course-rotation.

9 Rosamund Young, *The Secret Life of Cows*, Faber and Faber, London, 2017, p. 54.

10 Andy Bloomfield, 'The Spoonbill – A Holkham success story', 29 November 2018, https://www.holkham.co.uk/blog/post/the-spoonbill-a-holkham-success-story.

11 Richard F. Pywell et al., 'Wildlife-friendly farming increases crop yield: Evidence for ecological intensification', *Proceedings of the Royal Society B*, 7 October 2015, 282, https://royalsocietypublishing.org/doi/full/10.1098/rspb.2015.1740.

12 M.A. Pavao-Zuckerman, 'Soil Ecology', in *Encyclopedia of Ecology*, Academic Press, Cambridge, Massachusetts, 2008, https://www.sciencedirect.com/topics/agricultural-and-biological-sciences/soil-ecology/.

13 European Commission, *The Factory of Life: Why soil biodiversity is so important*, Office for Official Publications of the European Communities, Luxembourg, 2010, https://ec.europa.eu/environment/archives/soil/pdf/soil_biodiversity_brochure_en.pdf.

14 D.H. Wall and M.A. Knox, 'Soil Biodiversity', in *Reference Module in Earth Systems and Environmental Sciences*. Elsevier, Amsterdam,

2014, https://www.sciencedirect.com/topics/agricultural-and-biologi-cal-sciences/soil-biodiversity.

15 Bryan Griffiths, Chris McDonald and Mary-Jane Lawrie, 'Soil Biodiversity and Soil Health', Farm Advisory Service, Scotland, April 2019, Technical Note 721, https://www.fas.scot/downloads/tn721-soil-biodiversity-and-soil-health/.

16 WWF, 'Dishing the dirt on the secret life of soil', WWF, Washington DC, 5 December 2018, https://www.worldwildlife.org/stories/dishing-the-dirt-on-the-secret-life-of-soil.

17 John Crawford, personal communication, 2020.

18 Wikipedia, 'Earthworm', https://en.wikipedia.org/wiki/Earthworm.

19 European Commission, *The Factory of Life*.

20 Stephanie Pappas, 'Confirmed: The Soil Under Your Feet is Teeming with Life', *Live Science*, May 2016,
https://www.livescience.com/54862-soil-teeming-with-life.html.

21 Ibid.

22 M.A. Pavao-Zuckerman, 'Soil Ecology'.

23 FAO, 'Where Food Begins', http://www.fao.org/resources/infographics/infographics-details/en/c/285853/.

24 European Commission, *The Factory of Life*.

EPILOGUE

1 Matt McGrath, 'Biden: This will be "decisive decade" for tackling climate change,' (BBC News, Science & Environment, 22 April 2021), https://www.bbc.co.uk/news/science-environment-56837927.

2 Henry Zeffman and John Reynolds, 'Climate change summit: Veganism "will help Britain hit emissions targets"', (*Times*, 22 April 2021), https://www.thetimes.co.uk/article/climate-change-summit-this-is-the-year-to-get-serious-boris-johnson-tells-world-kkz76t5sv.

3 United Nations Secretary General Statement, 'Secretary-General's message on World Food Day', (UN New York, 16 October 2019), https://www.un.org/sg/en/content/sg/statement/2019-10-16/secretary-generals-message-world-food-day-scroll-down-for-french-version.

4 Bradshaw et al., 'Underestimating the Challenges of Avoiding a Ghastly Future', *Frontiers in Conservation Science*, 13 January 2021, 1,

p. 9, https://www.frontiersin.org/articles/10.3389/fcosc.2020.615419/full.

5 Helen Harwatt and Matthew N. Hayek, 'Eating away at climate change with negative emissions' Animal Law and Policy Program, Harvard Law School, 11 April 2019, https://animal.law.harvard.edu/wp-content/uploads/Eating-Away-at-Climate-Change-with-Negative-Emissions%E2%80%93%E2%80%93Harwatt-Hayek.pdf.

6 Daphne Ewing-Chow, 'This new food label will mainstream Whole Foods' biggest trend for 2020', Forbes, 20 December 2019, https://www.forbes.com/sites/daphneewingchow/2019/12/20/this-new-food-label-will-mainstream-whole-foods-biggest-trend-for-2020/?sh=53c77fb93933; The Ohio State University, CFAES on Sustainability, 'Dig the solution: How to offset 100 percent of all greenhouse gas emissions', (31 July 2015), https://u.osu.edu/sustainability/2015/07/31/dig-the-solution-how-to-offset-100-percent-of-all-greenhouse-gas-emissions/.

INDEX

A NOTE ON THE TYPE

The text of this book is set Adobe Garamond. It is one of several versions of Garamond based on the designs of Claude Garamond. It is thought that Garamond based his font on Bembo, cut in 1495 by Francesco Griffo in collaboration with the Italian printer Aldus Manutius. Garamond types were first used in books printed in Paris around 1532. Many of the present-day versions of this type are based on the Typi Academiae of Jean Jannon cut in Sedan in 1615.

Claude Garamond was born in Paris in 1480. He learned how to cut type from his father and by the age of fifteen he was able to fashion steel punches the size of a pica with great precision. At the age of sixty he was commissioned by King Francis I to design a Greek alphabet, and for this he was given the honourable title of Royal Type Founder. He died in 1561.

ABOUT COMPASSION IN WORLD FARMING

Compassion in World Farming is a powerful global movement dedicated to ending factory farming and radically changing our food systems to reduce reliance on animal protein, before it's too late.

Ending factory farming will bring better lives to billions of farm animals, save wildlife from extinction, improve our health, and leave a planet fit for future generations.

Factory farming is the single biggest cause of animal cruelty and a major driver of climate change so only by ending the needless suffering of farm animals can we hope to safeguard the future of our planet. By campaigning to end it, we are helping to shape an animal-friendly and nature-positive future for all life on this planet.

Find out more at www.ciwf.org

COMPASSION
in world farming
ciwf.org

Registered Charity Number 1095050